T0225758

SpringerBriefs in Applied Sciences and Technology

Nanoscience and Nanotechnology

Series editors

Hilmi Volkan Demir, Nanyang Technological University, Singapore, Singapore
Alexander O. Govorov, Department of Physics and Astronomy, Ohio University, Athens, OH, USA

Nanoscience and nanotechnology offer means to assemble and study superstructures, composed of nanocomponents such as nanocrystals and biomolecules, exhibiting interesting unique properties. Also, nanoscience and nanotechnology enable ways to make and explore design-based artificial structures that do not exist in nature such as metamaterials and metasurfaces. Furthermore, nanoscience and nanotechnology allow us to make and understand tightly confined quasi-zero-dimensional to two-dimensional quantum structures such as nanopalettes and graphene with unique electronic structures. For example, today by using a biomolecular linker, one can assemble crystalline nanoparticles and nanowires into complex surfaces or composite structures with new electronic and optical properties. The unique properties of these superstructures result from the chemical composition and physical arrangement of such nanocomponents (e.g., semiconductor nanocrystals, metal nanoparticles, and biomolecules). Interactions between these elements (donor and acceptor) may further enhance such properties of the resulting hybrid superstructures. One of the important mechanisms is excitonics (enabled through energy transfer of exciton-exciton coupling) and another one is plasmonics (enabled by plasmon-exciton coupling). Also, in such nanoengineered structures, the light-material interactions at the nanoscale can be modified and enhanced, giving rise to nanophotonic effects. These emerging topics of energy transfer, plasmonics, metastructuring and the like have now reached a level of wide-scale use and popularity that they are no longer the topics of a specialist, but now span the interests of all "end-users" of the new findings in these topics including those parties in biology, medicine, materials science and engineerings. Many technical books and reports have been published on individual topics in the specialized fields, and the existing literature have been typically written in a specialized manner for those in the field of interest (e.g., for only the physicists, only the chemists, etc.). However, currently there is no brief series available, which covers these topics in a way uniting all fields of interest including physics, chemistry, material science, biology, medicine, engineering, and the others. The proposed new series in "Nanoscience and Nanotechnology" uniquely supports this cross-sectional platform spanning all of these fields. The proposed briefs series is intended to target a diverse readership and to serve as an important reference for both the specialized and general audience. This is not possible to achieve under the series of an engineering field (for example, electrical engineering) or under the series of a technical field (for example, physics and applied physics), which would have been very intimidating for biologists, medical doctors, materials scientists, etc. The Briefs in NANOSCIENCE AND NANOTECHNOLOGY thus offers a great potential by itself, which will be interesting both for the specialists and the non-specialists. Indexed by SCOPUS.

More information about this series at http://www.springer.com/series/11713

Zi-Hui Zhang · Chunshuang Chu ·
Kangkai Tian · Yonghui Zhang

Deep Ultraviolet LEDs

Understanding the Low External Quantum
Efficiency

 Springer

Zi-Hui Zhang
School of Electronics and Information
Engineering, Institute of Micro-Nano
Photoelectron and Electromagnetic
Technology Innovation
Hebei University of Technology
Tianjin, Hebei, China

Key Laboratory of Electronic
Materials and Devices of Tianjin
Tianjin, China

Kangkai Tian
School of Electronics and Information
Engineering, Institute of Micro-Nano
Photoelectron and Electromagnetic
Technology Innovation
Hebei University of Technology
Tianjin, Hebei, China

Key Laboratory of Electronic
Materials and Devices of Tianjin
Tianjin, China

Chunshuang Chu
School of Electronics and Information
Engineering, Institute of Micro-Nano
Photoelectron and Electromagnetic
Technology Innovation
Hebei University of Technology
Tianjin, Hebei, China

Key Laboratory of Electronic
Materials and Devices of Tianjin
Tianjin, China

Yonghui Zhang
School of Electronics and Information
Engineering, Institute of Micro-Nano
Photoelectron and Electromagnetic
Technology Innovation
Hebei University of Technology
Tianjin, Hebei, China

Key Laboratory of Electronic
Materials and Devices of Tianjin
Tianjin, China

ISSN 2191-530X   ISSN 2191-5318 (electronic)
SpringerBriefs in Applied Sciences and Technology
ISSN 2196-1670   ISSN 2196-1689 (electronic)
Nanoscience and Nanotechnology
ISBN 978-981-13-6178-4   ISBN 978-981-13-6179-1 (eBook)
https://doi.org/10.1007/978-981-13-6179-1

Library of Congress Control Number: 2019930970

This Springer imprint is published by the registered company Springer Nature Singapore Pte Ltd.
The registered company address is: 152 Beach Road, #21-01/04 Gateway East, Singapore 189721, Singapore

Acknowledgements

We acknowledge the financial support by National Natural Science Foundation of China (Project Nos. 51502074, 61604051); Natural Science Foundation of Hebei Province (Project No. F2017202052); Natural Science Foundation of Tianjin City (Project No. 16JCYBJC16200); Program for Top 100 Innovative Talents in Colleges and Universities of Hebei Province (Project No. SLRC2017032); Program for 100-Talent-Plan of Hebei Province (Project No. E2016100010).

Contents

Abstract

AlGaN-based deep ultraviolet light-emitting diodes (DUV LEDs) are featured with small size, DC driving, no environmental contamination, etc., and they are now emerging as the excellent solid-state light source to replace the conventional mercury-based light tubes. Nevertheless, the DUV LEDs are currently affected by the poor external quantum efficiency (EQE), which is caused by the low internal quantum efficiency (IQE) and very unsatisfying light extraction efficiency (LEE). In this work, we will disclose the underlying mechanism for the low EQE and summarize the technologies that have been adopted so far for enhancing the EQE.

Chapter 1
Introduction

Abstract After the successful commercialization for InGaN/GaN blue light-emitting diodes that are used to generate white light, the development of AlGaN based deep ultraviolet light-emitting diodes (DUV LEDs) promises the complete replacement for mercury-based fluorescence light tubes and this guarantees that the Minamata Convention on Mercury can be fulfilled by the end of the year of 2020. Therefore, developing high-efficiency AlGaN based DUV LEDs is regarded as the next revolutionary event for solid-state lighting. In this book, we will review the current status and summarize the challenges for DUV LEDs. Meanwhile, we also suggest the research spots that are worth investigating for DUV LEDs.

As the solid-state light source, AlGaN-based deep ultraviolet light-emitting diodes (DUV LEDs) are energy-saving, DC driving, portable with very small size, contamination-free to the globe, which make DUV LEDs as very excellent candidate for water sterilization, air purification etc. [1]. Therefore, AlGaN-based DUV LEDs have attracted significant research efforts recently. However, at the current stage, the external quantum efficiency (EQE) for DUV LEDs is low, and most of the reported EQE is even lower than 10% for the devices with the peak emission wavelength shorter than 280 nm (see Fig. 1.1) [2], which EQE number cannot efficiently kill all the bacteria in the drinking water with high flow rate. As a result, before the massive penetration into the market to replace the conventional mercury based deep ultraviolet light source, it is essentially important to enhance the EQE for DUV LEDs. The EQE is co-affected by both the internal quantum efficiency (IQE) and the light extraction efficiency (LEE) [3–5].

As is well known, the IQE for III-nitride based LEDs is subject to the crystalline quality, the carrier injection efficiency, the current spreading, the self-heating effect, the polarization induced electric field within the [0001] oriented multiple quantum wells (MQWs), etc. [1, 3–6]. On the other hand, the TM-polarized optical photons for DUV LEDs make the mechanism of the light propagation unique, such that the TM-polarized light tends to escape from the sidewalls for the devices [7]. Meanwhile, the strong optical absorption for DUV photons arises from the passive layers with the smaller energy band gap and the absorptive contact metal further reduces the LEE [8]. Therefore, tremendous efforts shall be made to improve both the IQE and the LEE for

Z.-H. Zhang et al., *Deep Ultraviolet LEDs*, Nanoscience and Nanotechnology,
https://doi.org/10.1007/978-981-13-6179-1_1

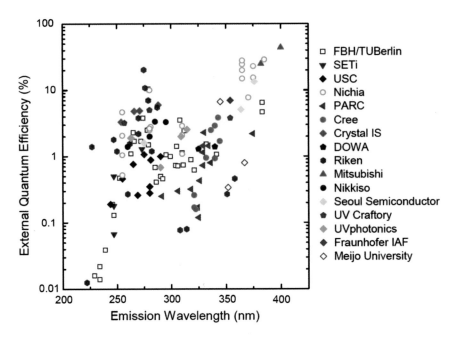

Fig. 1.1 Summary of the reported EQE for DUV LEDs. Reproduced from Ref. [2], with the permission of MDPI

DUV LEDs. This work will review the most recently reported methods to enhance the EQE for DUV LEDs. In the meantime, the underlying physical mechanism is also discussed.

References

1. Khan A, Balakrishnan K, Katona T (2008) Ultraviolet light-emitting diodes based on group three nitrides. Nat Photonics 2(2):77–84. https://doi.org/10.1038/nphoton.2007.293
2. Ding K, Avrutin V, Ozgur U, Morkoc H (2017) Status of growth of group III-nitride heterostructures for deep ultraviolet light-emitting diodes. Crystals 7(10):300. https://doi.org/10.3390/cryst7100300
3. Schubert EF (2006) Light-emitting diodes, 2nd edn. Cambridge University Press
4. Zhang Z-H, Zhang Y, Bi W, Demir HV, Sun XW (2016) On the internal quantum efficiency for InGaN/GaN light-emitting diodes grown on insulating substrates. Phys Status Solidi (a) 213(12):3078–3102. https://doi.org/10.1002/pssa.201600281
5. Nagasawa Y, Hirano A (2018) A review of AlGaN-based deep-ultraviolet light-emitting diodes on sapphire. Appl Sci (Basel) 8(8):1264. https://doi.org/10.3390/app8081264
6. Cho J, Schubert EF, Kim JK (2013) Efficiency droop in light-emitting diodes: challenges and countermeasures. Laser Photonics Rev 7(3):408–421. https://doi.org/10.1002/lpor.201200025
7. Nam KB, Li J, Nakarmi ML, Lin JY, Jiang HX (2004) Unique optical properties of AlGaN alloys and related ultraviolet emitters. Appl Phys Lett 84(25):5264–5266. https://doi.org/10.1063/1.1765208

8. Zhang YW, Jamal-Eddine Z, Akyol F, Bajaj S, Johnson JM, Calderon G, Allerman AA, Moseley MW, Armstrong AM, Hwang J, Rajan S (2018) Tunnel-injected sub 290 nm ultra-violet light emitting diodes with 2.8% external quantum efficiency. Appl Phys Lett 112(7):071107. https://doi.org/10.1063/1.5017045

Chapter 2
Increase the IQE by Improving the Crystalline Quality for DUV LEDs

Abstract The roadmap for AlGaN based DUV LEDs is similar to that for InGaN based visible LEDs, such that the success of achieving high crystalline-quality epilayers is the precondition for fabricating high-brightness DUV LEDs. This chapter will review the most adopted technologies for growing high-quality Al-rich AlGaN films, which is regarded as the milestone for making high-efficiency DUV LEDs.

There occur very huge lattice mismatch and the thermal mismatch when the Al-rich AlGaN layers are grown on the flat sapphire substrate [1, 2], which cause the high threading dislocation density (TDD) in the order of 10^{10}–10^{11} cm^{-2} [3]. According to the report by Khan et al., the absence of the localized states in the AlGaN based quantum wells further makes radiative recombination sensitive to the TDs for DUV LEDs [4], thus causing the IQE as low as 10% according to Fig. 2.1 [5, 6]. As a result, substantial efforts have to be made to reduce the TDD and correspondingly improve the IQE. For achieving that goal, nano-patterned sapphire substrates have been proposed and fabricated. The growth of the Al-rich AlGaN layer on the nano-patterned sapphire substrate originates from the fact that the low mobility for the Al adatoms causes a long coalescence time for the AlN buffer layer [7–9]. At the current stage, the epitaxial growth on nano-patterned substrates is deemed as the most reliable technique for growing high-quality AlGaN based DUV LEDs. The advantage of the growing Al-rich AlGaN based DUV LEDs on nano-patterned substrates has been studied and shows the advantage in suppressing the TDD and reducing the Shockley-Read-Hall (SRH) lifetime [10–12]. According to the report by Zhang et al., the edge and screw dislocation density can be reduced to 6.3×10^7 cm^{-2} and 3.2×10^8 cm^{-2}, respectively [12], which number can predict the intrinsic IQE as high as 70% according to Fig. 2.1 [5]. Here, the intrinsic IQE is measured by using the low-temperature photoluminescence method, which does not get the carrier injection involved. However, if the growth cost is not considered, then the TDD can be remarkably reduced to 10^3–10^5 cm^{-2}, which number is obtained by growing Al-rich AlGaN layers on the free-standing AlN substrate [13–19]. The DUV LED grown

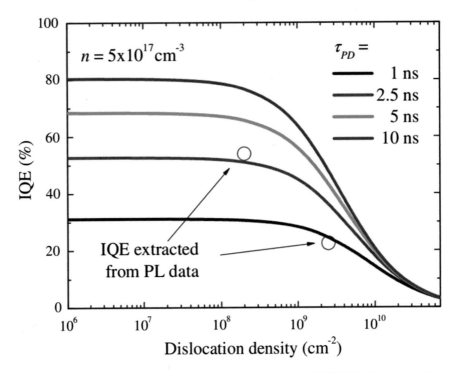

Fig. 2.1 Calculated IQE in terms of different dislocation density for DUV LEDs. Reproduced from Ref. [5], with the permission of IOP Publishing

on AlN free-standing substrate can make the intrinsic IQE very close to 100%. Ref. [20] has comprehensively summarized the most recently reported TDD values that are obtained by different growth techniques on various substrates, by combing Ref. [20] and Fig. 2.1, one can easily estimate the intrinsic IQE for DUV LEDs grown by different technique.

References

1. Imura M, Nakano K, Narita G, Fujimoto N, Okada N, Balakrishnan K, Iwaya M, Kamiyama S, Amano H, Akasaki I, Noro T, Takagi T, Bandoh A (2007) Epitaxial lateral overgrowth of AlN on trench-patterned AlN layers. J Cryst Growth 298:257–260. https://doi.org/10.1016/j.jcrysgro.2006.10.043
2. Ambacher O (1998) Growth and applications of Group III-nitrides. J Phys D Appl Phys 31(20):2653
3. Masataka I, Kiyotaka N, Naoki F, Narihito O, Krishnan B, Motoaki I, Satoshi K, Hiroshi A, Isamu A, Tadashi N, Takashi T, Akira B (2007) Dislocations in AlN epilayers grown on sapphire substrate by high-temperature metal-organic vapor phase epitaxy. Jpn J Appl Phys 46(4A):1458–1462. https://doi.org/10.1143/JJAP.46.1458

4. Khan A, Balakrishnan K, Katona T (2008) Ultraviolet light-emitting diodes based on group three nitrides. Nat Photonics 2(2):77–84. https://doi.org/10.1038/nphoton.2007.293
5. Shatalov M, Sun W, Lunev A, Hu X, Dobrinsky A, Bilenko Y, Yang J, Shur M, Gaska R, Moe C, Garrett G, Wraback M (2012) AlGaN deep-ultraviolet light-emitting diodes with external quantum efficiency above 10%. Appl Phys Express 5(8):082101. https://doi.org/10.1143/APEX.5.082101
6. Kneissl M, Kolbe T, Chua C, Kueller V, Lobo N, Stellmach J, Knauer A, Rodriguez H, Einfeldt S, Yang Z, Johnson NM, Weyers M (2011) Advances in group III-nitride-based deep UV light-emitting diode technology. Semicond Sci Technol 26(1):014036. https://doi.org/10.1088/0268-1242/26/1/014036
7. Hirayama H, Norimatsu J, Noguchi N, Fujikawa S, Takano T, Tsubaki K, Kamata N (2009) Milliwatt power 270 nm-band AlGaN deep-UV LEDs fabricated on ELO-AlN templates. Phys Status Solidi C 6:S474–S477. https://doi.org/10.1002/pssc.200880959
8. Vinod A, Qhalid F, Monirul I, Thomas K, Balakrishnan K, Asif K (2007) Robust 290 nm emission light emitting diodes over pulsed laterally overgrown AlN. Jpn J Appl Phys 46(36–40):L877–L879. https://doi.org/10.1143/JJAP.46.L877
9. Kim M, Fujita T, Fukahori S, Inazu T, Pernot C, Nagasawa Y, Hirano A, Ippommatsu M, Iwaya M, Takeuchi T, Kamiyama S, Yamaguchi M, Honda Y, Amano H, Akasaki I (2011) AlGaN-based ultraviolet light-emitting diodes fabricated on patterned sapphire substrates. Appl Phys Express 4(9):092102. https://doi.org/10.1143/APEX.4.092102
10. Dong P, Yan J, Wang J, Zhang Y, Geng C, Wei T, Cong P, Zhang Y, Zeng J, Tian Y, Sun L, Yan Q, Li J, Fan S, Qin Z (2013) 282-nm AlGaN-based deep ultraviolet light-emitting diodes with improved performance on nano-patterned sapphire substrates. Appl Phys Lett 102(24):241113. https://doi.org/10.1063/1.4812237
11. Dong P, Yan J, Zhang Y, Wang J, Zeng J, Geng C, Cong P, Sun L, Wei T, Zhao L, Yan Q, He C, Qin Z, Li J (2014) AlGaN-based deep ultraviolet light-emitting diodes grown on nano-patterned sapphire substrates with significant improvement in internal quantum efficiency. J Cryst Growth 395:9–13. https://doi.org/10.1016/j.jcrysgro.2014.02.039
12. Zhang L, Xu F, Wang J, He C, Guo W, Wang M, Sheng B, Lu L, Qin Z, Wang X, Shen B (2016) High-quality AlN epitaxy on nano-patterned sapphire substrates prepared by nano-imprint lithography. Sci Rep 6:35934. https://doi.org/10.1038/srep35934
13. Available: http://www.hexatechinc.com/aln-wafer-sales.html
14. Hartmann C, Wollweber J, Dittmar A, Irmscher K, Kwasniewski A, Langhans F, Neugut T, Bickermann M (2013) Preparation of bulk AlN seeds by spontaneous nucleation of freestanding crystals. Jpn J Appl Phys 52(8):UNSP 08JA06. https://doi.org/10.7567/jjap.52.08ja06
15. Herro ZG, Zhuang D, Schlesser R, Sitar Z (2010) Growth of AlN single crystalline boules. J Cryst Growth 312(18):2519–2521. https://doi.org/10.1016/j.jcrysgro.2010.04.005
16. Dalmau R, Moody B, Xie J, Collazo R, Sitar Z (2011) Characterization of dislocation arrays in AlN single crystals grown by PVT. Phys Status Solidi a Appl Mater Sci 208(7):1545–1547. https://doi.org/10.1002/pssa.201000957
17. Sumathi RR (2013) Bulk AlN single crystal growth on foreign substrate and preparation of free-standing native seeds. CrystEngComm 15(12):2232–2240. https://doi.org/10.1039/c2ce26599k
18. Mokhov E, Izmaylova I, Kazarova O, Wolfson A, Nagalyuk S, Litvin D, Vasiliev A, Helava H, Makarov Y (2013) Specific features of sublimation growth of bulk AlN crystals on SiC wafers. Phys Status Solidi C 10(3):445–448. https://doi.org/10.1002/pssc.201200638
19. Bondokov RT, Mueller SG, Morgan KE, Slack GA, Schujman S, Wood MC, Smart JA, Schowalter LJ (2008) Large-area AlN substrates for electronic applications: an industrial perspective. J Cryst Growth 310(17):4020–4026. https://doi.org/10.1016/j.jcrysgro.2008.06.032
20. Li DB, Jiang K, Sun XJ, Guo CL (2018) AlGaN photonics: recent advances in materials and ultraviolet devices. Adv Opt Photonics 10(1):43–110. https://doi.org/10.1364/AOP.10.000043

Chapter 3
Improve the Current Spreading for DUV LEDs

Abstract After the crystalline quality for Al-rich AlGaN layer is significantly improved, it is then the time to design novel DUV LED structures. DUV LEDs are driven electrically which get carrier transport and current injection involved. One of the challenges is the current crowding effect, which easily occurs in the DUV LEDs. Hence, it is important to show people physical images on the underlying reason for the current crowding and the solution proposals for current spreading.

The non-active layers for DUV LEDs have the energy band smaller than the energy of the DUV photons, and therefore the EQE for DUV LEDs is strongly affected by the optical absorption [1, 2]. The non-active AlGaN layers are also Al-rich, which makes the ionization efficiency for dopants low and leads to the bad electrical conductivity [3]. On the other hand, most of the DUV LEDs are growing on the insulating sapphire substrate. Thus flip-chip structures are utilized for better light extraction efficiency. Nevertheless, the flip-chip DUV LEDs require the n-electrode and the p-electrode on the same side. As a result, significant nonuniform current distribution readily occurs for DUV LEDs [4, 5]. Currently, reports on the techniques to improve the currents spreading for DUV LEDs are rare. Attention has been made to optimize the pattern for the p-electrode [6, 7], and the optimized p-electrode can reduce the current crowding effect, suppress the self-heating effect and improve the wall-plug efficiency (WPE). Hrong et al. propose to deposit the Zinc gallate (ZnGa2O4; ZGO) thin film serving as the current spreading layer for DUV LEDs [8]. The advantages of the ZGO include very high transmittance in the 280 nm range and excellent electrical conductivity. Another design concern regarding the current spreading is the chip size for DUV LEDs, and according to the report by Kim et al. [9], a circular chip with properly small chip size is helpful to enhance the current spreading effect. However, we suggest improving the current spreading effect by epitaxially growing in situ current spreading layer in the DUV LED architectures, which can remarkably save the budget and the difficulty for fabricating the chips.

Z.-H. Zhang et al., *Deep Ultraviolet LEDs*, Nanoscience and Nanotechnology,
https://doi.org/10.1007/978-981-13-6179-1_3

References

1. Ryu H-Y, Choi I-G, Choi H-S, Shim J-I (2013) Investigation of light extraction efficiency in AlGaN deep-ultraviolet light-emitting diodes. Appl Phys Express 6(6):062101. https://doi.org/10.7567/APEX.6.062101
2. Maeda N, Hirayama H (2013) Realization of high-efficiency deep-UV LEDs using transparent p-AlGaN contact layer. Phys Status Solidi C 10(11):1521–1524. https://doi.org/10.1002/pssc.201300278
3. Katsuragawa M, Sota S, Komori M, Anbe C, Takeuchi T, Sakai H, Amano H, Akasaki I (1998) Thermal ionization energy of Si and Mg in AlGaN. J Cryst Growth 189:528–531. https://doi.org/10.1016/S0022-0248(98)00345-5
4. Schubert EF (2006) Light-emitting diodes, 2nd edn. Cambridge University Press
5. Guo X, Schubert EF (2001) Current crowding in GaN/InGaN light emitting diodes on insulating substrates. J Appl Phys 90(8):4191–4195. https://doi.org/10.1063/1.1403665
6. Hao GD, Taniguchi M, Tamari N, Inoue S (2016) Enhanced wall-plug efficiency in AlGaN-based deep-ultraviolet light-emitting diodes with uniform current spreading p-electrode structures. J Phys D-Appl Phys 49(23):235101. https://doi.org/10.1088/0022-3727/49/23/235101
7. Hao GD, Taniguchi M, Tamari N, Inoue S (2018) Current crowding and self-heating effects in AlGaN-based flip-chip deep-ultraviolet light-emitting diodes. J Phys D-Appl Phys 51(3):035103. https://doi.org/10.1088/1361-6463/aa9e0e
8. Hrong RH, Zeng YY, Wang WK, Tsai CL, Fu YK, Kuo WH (2017) Transparent electrode design for AlGaN deep-ultraviolet light-emitting diodes. Opt Express 25(25):32206–32213. https://doi.org/10.1364/OE.25.032206
9. Kim KH, Fan ZY, Khizar M, Nakarmi ML, Lin JY, Jiang HX (2004) AlGaN-based ultraviolet light-emitting diodes grown on AlN epilayers. Appl Phys Lett 85(20):4777–4779. https://doi.org/10.1063/1.1819506

Chapter 4
Improve the Hole Injection to Enhance the IQE for DUV LEDs

Abstract The very low doping efficiency for the p-type Al-rich AlGaN layers indicates that the hole injection capability for DUV LEDs can be poor. Therefore, we ought to investigate the approaches to enable high-efficiency hole injection. In this chapter, we propose novel DUV LED architectures to make "hot" holes, increase the hole concentration in the p-type layer, and reduce the hole blocking effect that arises from the p-type electron blocking layer (p-EBL).

Although the very high intrinsic IQE can be obtained as long as the TDD can be reduced to a decent level, the DUV LEDs are electrically driven. Hence the IQE is strongly associated with the carrier injection. The hole injection layer for AlGaN based DUV LEDs normally comprises the p-type electron blocking layer (p-EBL)/p-AlGaN/p-GaN structure. This illustrates that the Mg ionization efficiency be even smaller than 1% at room temperature, and hence the free hole concentration can be lower than 10^{17} cm^{-3} [1]. Moreover, the holes will experience two energy barriers when traveling through the p-EBL/p-AlGaN/p-GaN structure, which further retards the hole transport. Currently, one approach to enhance the hole injection efficiency is to increase the Mg doping efficiency for the p-AlGaN layer. The enhanced Mg doping efficiency can be achieved by using the three-dimensional hole gas (3DHG) [2, 3], Mg-delta doping [4–7] and the indium-surfactant-assisted Mg-delta doping [8]. The indium-surfactant-assisted Mg-delta doping method can increase the free hole concentration up to 4.75×10^{18} cm^{-3} for the p-GaN layer according to the report by Chen et al. [8]. Besides, the superlattice structure is another alternative to enhance the Mg ionization coefficient, such that the polarization induced electric field in the superlattice structure triggers the Poole-Frenkel effect [9–13]. Another proposal to increase the hole transport across the p-type layer is to suppress the barrier height for holes, which is doable by adopting the p-AlGaN layer with the multiple stair-cased AlN compositions or grading the AlN composition [14, 15].

© The Author(s), under exclusive license to Springer Nature Singapore Pte Ltd. 2019   11
Z.-H. Zhang et al., *Deep Ultraviolet LEDs*, Nanoscience and Nanotechnology,
https://doi.org/10.1007/978-981-13-6179-1_4

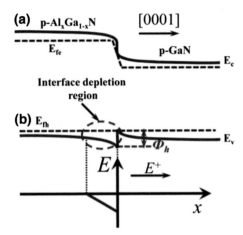

Fig. 4.1 a Schematic energy band diagram for the electric-field reservoir, in which there exists an interface depletion region in the p-Al$_x$Ga$_{1-x}$N layer, **b** schematic electric field profile in the interface depletion region. Here, E_c, E_v, E_{fe}, E_{fh} and Φ_h denote the conduction band, the valence band, the quasi-Fermi level for electrons, the quasi-Fermi level for holes, and the effective valence band barrier height for the p-Al$_x$Ga$_{1-x}$N layer side. Note, the alignment of the quasi-Fermi level for holes sketches that there exists a hole depletion region in the p-Al$_x$Ga$_{1-x}$N side for the p-Al$_x$Ga$_{1-x}$N/p-GaN interface, i.e., interface depletion region. Reproduced from Ref. [18], with the permission of Optical Society of America

4.1   Make Holes "Hot" for DUV LEDs

As is well known, the hole transport across the p-EBL strongly depends on the hole energy [16, 17]. Unfortunately, the mobility for holes is smaller than that for electrons, which represents that the drift velocity and the thermal energy for holes have to be enhanced from another prospect, i.e., playing with the electric field. For that purpose, we propose the electric field reservoir (EFR) which proves to be effective in increasing the hole energy for DUV LEDs [18]. The schematic energy band diagram of the EFR for DUV LEDs is shown in Fig. 4.1a. The EFR is a p-Al$_x$Ga$_{1-x}$N/p-GaN heterojunction, which has a very pronounced impact on the hole injection. There exists the build-in electric field in the p-Al$_x$Ga$_{1-x}$N/p-GaN interface according to Fig. 4.1b. Fortunately, the electric field is of the same direction as the one that is generated by the external bias, and this helps to increase the drift velocity and the corresponding thermal energy for holes. On the other hand, if the build-in electric field penetrates the whole p-Al$_x$Ga$_{1-x}$N layer, then the holes will be substantially depleted. As a result, the EFR has to increase the hole energy without significantly depleting the holes.

To better clarify the effect of the EFR on the hole injection and the optical power for DUV LEDs, five devices are designed and investigated. The structural information for different p-EBL/p-Al$_x$Ga$_{1-x}$N architectures are demonstrated in Table 4.1. The electric field profiles within the p-Al$_x$Ga$_{1-x}$N/p-GaN layers for different DUV LEDs

Table 4.1 Device structures for the studied DUV LEDs

Devices	p-Al$_x$Ga$_{1-x}$N	Φ_h(meV)	p-Al$_y$Ga$_{1-y}$N	Work (meV)
Original device	p-GaN (50 nm)	0	p-Al$_{0.68}$Ga$_{0.32}$N EBL (10 nm)	−277.50
Reference device	p-Al$_{0.49}$Ga$_{0.51}$N (50 nm)	583.00	p-Al$_{0.68}$Ga$_{0.32}$N EBL (10 nm)	−7454.70
Device 1 (D1)	p-Al$_{0.49}$Ga$_{0.51}$N (50 nm)	460.00	p-Al$_{0.60}$Ga$_{0.40}$N EBL (10 nm)	−5456.10
Device 2 (D2)	p-Al$_{0.40}$Ga$_{60}$N (50 nm)	322.00	p-Al$_{0.68}$Ga$_{0.32}$N EBL (10 nm)	−381.97
Device 3 (D3)	P-Al$_{0.30}$Ga$_{0.70}$N (50 nm)	238.00	p-Al$_{0.68}$Ga$_{0.32}$N EBL (10 nm)	−365.72

Reproduced from Ref. [18], with the permission of Optical Society of America

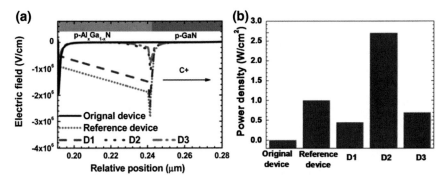

Fig. 4.2 **a** Electric field profiles within the p-type regions, **b** optical power density for the devices in Table 4.1. Fig. **a** is reproduced from Ref. [18], with the permission of Optical Society of America

are presented in Fig. 4.2a, from which we can see that the reference device and D1 have very strong electric field intensity and the p-Al$_x$Ga$_{1-x}$N layers are fully depleted. The hole depletion effect becomes small for D2 and D3. The energy that the holes obtain can be calculated by following $W = e \times \int_0^l E_{field} \times dx$, where e, l, E_{field} denote the unit electronic charge, the integration range and the electric field, respectively. The results are shown in Table 4.1. Note, the "-" sign means that the holes receive energy from the electric field. Although Reference device and D1 have obtained very big energy from the EFR, the significant hole depletion effect in the p-Al$_x$Ga$_{1-x}$N layer sacrifices the hole injection, and therefore the optical power density is not the strongest [see Fig. 4.2b]. Although the hole energy for D2 is not as large as that for Reference device and D1, the hole concentration in the p-Al$_x$Ga$_{1-x}$N layer is higher. As the result, D2 produces the largest optical power density according to Fig. 4.2b. We then will explain Fig. 4.2b in detail as follows.

Figure 4.3a shows the numerically calculated EQE and the optical power density in terms of the injection density, which is consistent with Fig. 4.2b. Figure 4.3a is

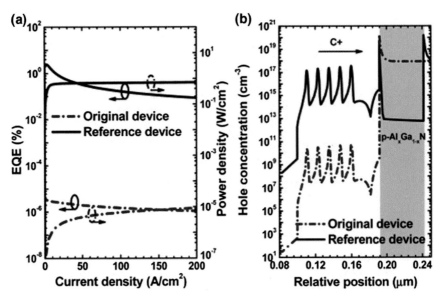

Fig. 4.3 a Numerically calculated EQE and optical power density, **b** hole concentration profiles in the MQWs and the p-Al$_x$Ga$_{1-x}$N layer for the original device and the reference device. Fig. **b** is calculated at the current density of 150 A/cm^2. Figures are reproduced from Ref. [18], with the permission of Optical Society of America

presented in semilog scale, and we can see that the enhancement of the EQE is significant, which is large as 10^5. Figure 4.3b shows the hole concentration profiles in the MQW region and in the p-Al$_x$Ga$_{1-x}$N layer for the original device and the reference device. It shows the very strong hole depletions in the p-Al$_x$Ga$_{1-x}$N layer for the reference device, which agrees well with our discussions for Fig. 4.2a. Nevertheless, according to Table 4.1, we can get that the holes are able to be substantially accelerated for the reference device than that for the original device. Therefore, the hole injection efficiency for the reference device is much stronger than that for the original device, which can be readily obtained by looking into the hole concentration profiles in the MQWs for the two devices. The hole concentration level for the reference device is improved by 10^7 according to our calculations. Investigations into Fig. 4.3a and b conclude that the EFR structure can increase the hole energy by making holes "hot". "Hot" holes can be more efficiently injected into the MQWs.

Figure 4.4a and b demonstrate the calculated EQE, the optical power density and the hole concentrations profiles for the reference device and Device 1, respectively. Table 4.1 shows that the holes can obtain less energy from Device 1, which explains the smaller EQE for Device 1 than that for the reference device. On the other hand, the difference for the hole energy between the reference device and Device 1 is smaller than that between the original device and the reference device. Therefore, the EQE enhancement in Fig. 4.4a is not as big as that in Fig. 4.3a. The same conclusion can also be made for Fig. 4.4b when compared with Fig. 4.3b, which further proves that

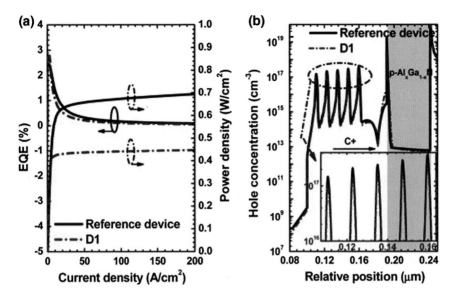

Fig. 4.4 **a** Numerically calculated EQE and optical power density, **b** hole concentration profiles in the MQWs and the p-Al$_x$Ga$_{1-x}$N layer for the reference device and Device 1. Fig. **b** is calculated at the current density of 150 A/cm^2. Figures are reproduced from Ref. [18], with the permission of Optical Society of America

the hole injection is strongly subject to the hole energy. However, we can also get that, in spite of the improvement when compared to the original device and Device 1, the EQE value for the reference device is as low as 1%. The very low EQE indicates that the structure for the reference device is not fully optimized, and the EFR needs further exploration. We believe that the hole injection for the reference device can be further promoted if the hole depletion effect within the p-Al$_x$Ga$_{1-x}$N layer can be properly suppressed.

The electric field profiles in Fig. 4.2a indicate that the hole depletion effect in the p-Al$_x$Ga$_{1-x}$N layer may be suppressed for Devices 2 and 3. Therefore, comparisons are conducted among the reference device, Devices 2 and 3 in Fig. 4.5a and b. We firstly show the calculated EQE and the optical power density as a function of the injection current density in Fig. 4.5a, which shows that the strongest EQE can be obtained from Device 2. The EQE for Device 2 can overtake the EQE for the reference device when the current density exceeds 50 A/cm^2 according to Fig. 4.5a. To interpret the origin for the improved EQE for Device 2, we then present the hole concentration profiles for the three investigated devices in Fig. 4.5b. We can see that the holes are less depleted in the p-Al$_x$Ga$_{1-x}$N layer for Devices 2 and 3, which therefore increases the hole concentration and promotes the hole injection. Comparison between the hole concentration profiles between Devices 2 and 3 shows that Device 2 has the higher hole concentration than Device 3, which is attributed to the lager hole energy according to Table 4.1. However, when we look into the hole concentration profiles in

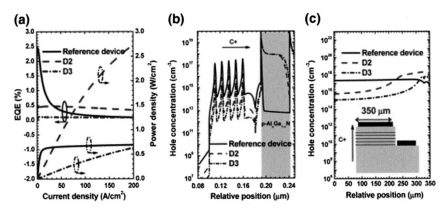

Fig. 4.5 **a** Numerically calculated EQE and optical power density, **b** hole concentration profiles in the MQWs and the p-Al$_x$Ga$_{1-x}$N layer, and **c** lateral hole concentration profiles in the p-Al$_x$Ga$_{1-x}$N layer for the reference device, Devices 2 and 3. Fig. **b** and **c** are calculated at the current density of 150 A/cm^2. Figures are reproduced from Ref. [18], with the permission of Optical Society of America

the MQWs for the three devices, we find that the hole concentration level for Devices 2 and 3 are lower than that for the reference device. Then, we realize that the hole concentration in the p-Al$_x$Ga$_{1-x}$N layer influences the electric conductivity and the current spreading. Meanwhile, the flip-chip DUV LEDs have both the p-electrode and the n-electrode on the same side, which further makes the lateral current more sensitive to the hole concentration in the p-Al$_x$Ga$_{1-x}$N layer as discussed in Sect. 4.3. We then present the lateral hole concentration profiles in the p-Al$_x$Ga$_{1-x}$N layer for the reference device, Devices 2 and 3 in Fig. 4.5c, which illustrates that the holes are more crowded at the right mesa edge. Nevertheless, the overall hole concentration for Device 2 is the highest among the three devices. Therefore the strongest optical power density is produced by Device 2 in Fig. 4.5a. Figure 4.5c implies the EQE for DUV LEDs can be further promoted if the current spreading can be improved, and this also indicates the importance of improving the current spreading for DUV LEDs.

The hole concentration and the electric field are directly linked with the current-voltage characteristics. Therefore, we calculate and present the relationship between the current and the voltage for the studied DUV LEDs. Figure 4.6a compares the current-voltage characteristics for the original device and the reference device, which illustrates that the forward voltage has been significant increase for the reference device. Although the original device consumes the small forward voltage, the very low EQE [see Fig. 4.3a] makes it impossible for the realistic usage. Currently, most of the DUV LEDs utilize the similar p-type hole injection layer as in the reference device, i.e., p-EBL/p-Al$_x$Ga$_{1-x}$N/p-GaN structure, which indicates the urgency in reducing the forward voltage and the understanding the device physics for the p-EBL/p-Al$_x$Ga$_{1-x}$N/p-GaN structured hole injection layer. The comparison between the reference device and Device 1 in Fig. 4.6b implies the slight reduction of the

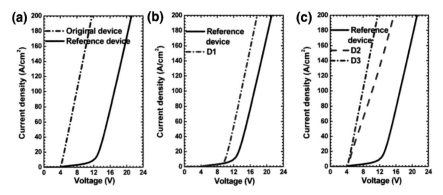

Fig. 4.6 Current-voltage characteristics for **a** the original device and the reference device, **b** the reference device and Device 1, **c** the reference device, Devices 2 and 3. Figures are reproduced from Ref. [18], with the permission of Optical Society of America

forward voltage can be obtained by reducing the AlN composition for the p-EBL layer. The reduced AlN composition decreases the electric field in the p-Al_xGa_{1-x}N layer according to Fig. 4.2a. However, the low AlN composition in the p-EBL may cause the poor electron injection efficiency. Then, we have to reduce the electric field in the p-Al_xGa_{1-x}N layer by utilizing the alternatives, i.e., reducing the AlN composition for the p-Al_xGa_{1-x}N layer. Therefore, we next compare the current-voltage characteristic for the reference device, Devices 2 and 3 in Fig. 4.6c. The forward voltages for Devices 2 and 3 have been largely reduced. If we refer to Fig. 4.2a, we can see that the electric field in the p-Al_xGa_{1-x}N layer for Devices 2 and 3 is not as strong as that for the reference device and Device 1. Once the electric field becomes weak, then the hole depletion effect in the p-Al_xGa_{1-x}N layer is also suppressed. Therefore, the forward voltage for Devices 2 and 3 can be remarkably reduced. Although Device 3 has the smallest forward voltage, the holes can get the least energy, and the increased valence band offset between the p-EBL and the p-Al_xGa_{1-x}N layer further hinders the hole injection. Therefore, we speculate that Device 2 can be the best design in our case.

We then calculate and present the wall-plug-efficiency in terms of the injection current density for the investigated DUV LEDs in Fig. 4.7a–c. Figure 4.7a shows that the reference device possesses the higher wall-plug-efficiency than the original device. The wall-plug-efficiency for Device 1 is comparable to the reference device according to Fig. 4.7b. Figure 4.7c illustrates that Device 2 produces the best wall-plug-efficiency. Moreover, we also know that Device 2 shows the highest EQE [see Fig. 4.5a] and the reduced forward voltage [see Fig. 4.6c]. Therefore, the most optimized structure is Device 2 in our case, and this agrees well with our analysis previously.

Hence, although the p-EBL/p-Al_xGa_{1-x}N/p-GaN structure has been widely used for AlGaN based DUV LEDs, the physical mechanism on the hole injection for the p-EBL/p-Al_xGa_{1-x}N/p-GaN structure is not clear till now. We find that by properly

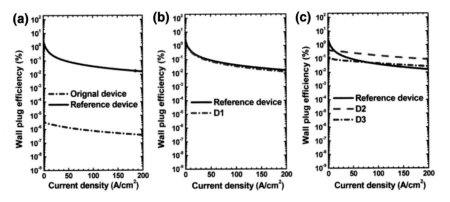

Fig. 4.7 Wall-plug-efficiency as a function of the injection current density for **a** the original device and the reference device, **b** the reference device and Device 1, **c** the reference device, Devices 2 and 3. Figures are reproduced from Ref. [18], with the permission of Optical Society of America

designing the p-EBL/p-Al$_x$Ga$_{1-x}$N/p-GaN structure, one can make "hot" holes without sacrificing the hole concentration in the p-EBL/p-Al$_x$Ga$_{1-x}$N/p-GaN structure. The advantage of the p-EBL/p-Al$_x$Ga$_{1-x}$N/p-GaN structure is that the electric field always exists at the p-Al$_x$Ga$_{1-x}$N/p-GaN interface without being screened by the free carriers. As a result, the holes can continuously obtain the energy by making themselves "hot". However, an unoptimized p-EBL/p-Al$_x$Ga$_{1-x}$N/p-GaN structure may significantly deplete the holes, which therefore leads to the insufficient hole injection and the very high forward voltage. Therefore, the findings here provide the additional understanding and enrich the device physics for AlGaN based DUV LEDs.

4.2 Superlattice p-EBL to Improve the Hole Injection Efficiency

Another obstacle for the hole injection arises from the p-EBL [19]. One promising way to suppress the hole blocking effect is to increase the Mg doping efficiency for the p-EBL, which, as has been mentioned previously, can be realized by utilizing superlattice structure [9–13]. The increased Mg doping efficiency and the corresponding improved hole concentration in the p-EBL can reduce the valence band barrier height [20]. Besides the Mg doping concentration, Kolbe et al. suggest that the AlN composition for the p-EBL shall also be fully optimized [21]. Recently, we have proposed a p-AlGaN/p-AlGaN superlattice EBL structure for DUV LEDs. The AlN composition for the proposed superlattice EBL is specially designed in the way that the superlattice loop starts from the thin p-AlGaN layer with a lower AlN composition. The detailed structure information for the p-EBLs is demonstrated in

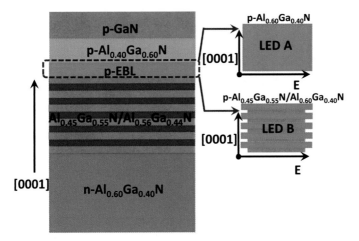

Fig. 4.8 Schematic structures for DUV LEDs with p-$Al_{0.60}Ga_{0.40}N$ EBL (LED A) and p-$Al_{0.60}Ga_{0.40}N$/p-$Al_{0.45}Ga_{0.55}N$ superlattice EBL (LED B). Reproduced from Ref. [13], with the permission of Springer

Fig. 4.8. LED A has the conventional p-$Al_{0.60}Ga_{0.40}N$ EBL and LED B has the p-$Al_{0.60}Ga_{0.40}N$/p-$Al_{0.45}Ga_{0.55}N$ superlattice EBL. As has been mentioned, the superlattice p-EBL initiates from the thin p-$Al_{0.45}Ga_{0.55}N$ layer, since by doing so, the $Al_{0.56}Ga_{0.44}N$/p-$Al_{0.45}Ga_{0.55}N$ interface can have the negative polarization induced sheet charges, which are very useful in increasing the electron blocking effect by the last quantum barrier.

We demonstrate the electroluminescence (EL) spectra for LEDs A and B in Fig. 4.9a, which shows that the EL intensity for LED B is stronger than that for LED A at the tested current density levels. With the EL spectra at different current density levels, we are able to get the EQE and the optical power density as a function of the injection current density as shown in Fig. 4.9b. When we compare LEDs A and B, the experimental enhancement is ~90% and the efficiency droop is reduced from 24 to 4% at the current density of 110 A/cm^2. Meanwhile, Fig. 4.9c presents the numerically calculated EQE and optical power density at different injection current density levels. Figure 4.9c has numerically reproduced Fig. 4.9b, such that the EQE has been remarkably improved and the efficiency droop has been substantially suppressed. The excellent agreement between the experimentally measured results and the numerically calculated ones indicates that the models and the parameters that we set in our numerical calculations are reasonable.

To probe the origin for the enhanced optical power for LED B, we calculate and show the hole concentration profiles for LEDs A and B in Fig. 4.10a. We can clearly see that the hole concentration level in the MQWs for LED B is much higher than that for LED A. The enhanced hole concentration translates to the improved radiative recombination rate in the MQWs. Figure 4.10b presents the hole concentration profiles in the p-EBLs and the p-$Al_{0.40}Ga_{0.60}N$ layers, which show that the superlattice p-EBL possesses the higher overall hole concentration. Moreover, the hole concentration at the interface of the superlattice p-EBL and the p-$Al_{0.40}Ga_{0.60}N$ layer

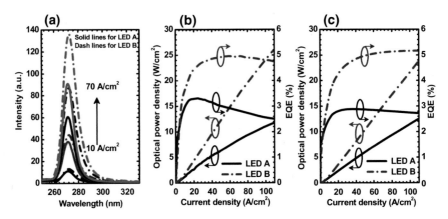

Fig. 4.9 **a** Measured EL spectra, **b** measured EQE and optical power density, **c** calculated EQE and optical power density. Reproduced from Ref. [13], with the permission of Springer

for LED B decreases when compared to that for LED A. The decreased hole concentration level at the p-EBL/p-Al$_{0.40}$Ga$_{0.60}$N interface well indicates the reduced hole blocking effect by the superlattice p-EBL. As we have mentioned earlier, the other advantage of this design is the reduced electron leakage. We show the EL spectra in the semilog scale in Fig. 4.10c, and we can see that the parasitic emission in the p-GaN layer for LED B produces the smaller intensity than that for LED A. The reduced intensity for the parasitic emission is ascribed to the reduced electron leakage. Figure 4.10d demonstrates the calculated electron concentration profiles in the p-EBLs and the p-Al$_{0.40}$Ga$_{0.60}$N layers for LEDs A and B, and it shows the suppressed electron leakage current by adopting the proposed superlattice p-EBL. The calculated results in Fig. 4.10d agree well with the speculations in Fig. 4.10c.

We calculate and present the energy band diagrams in the vicinity of the p-EBLs for LEDs A and B in Fig. 4.11a and b, respectively. The hole injection is strongly affected by the effective valence band barrier height for the p-EBL, which is represented as $\varnothing_h$. As has been shown in Fig. 4.10b, the superlattice p-EBL increases the hole concentration which enables a reduced $\varnothing_h$ for the p-EBL. The calculated values for $\varnothing_h$ are ~324 and ~281 meV for LEDs A and B, respectively at the current density of 50 A/cm^2. The strong polarization induced positive charges at the last quantum barrier/p-EBL interface can significantly attract electrons, giving rise to the high local electron concentration. The high local electron concentration can reduce the effective conduction band barrier height (i.e., $\varnothing_e$). The effective conduction band barrier height of the p-EBL is ~295 meV for LED A. The conduction band of the last quantum barrier for LED B is tilted upwards which helps to reduce the electron accumulation in the last quantum barrier. Therefore, the superlattice p-EBL increases the effective conduction band barrier height to ~391 meV. Figure 4.11b demonstrates the experimentally measured current density in terms of the applied voltage. The forward voltage for LED B is reduced when compared to LED A, which is attributed to the improved hole injection and the enhanced radiative recombination rate in the MQW region.

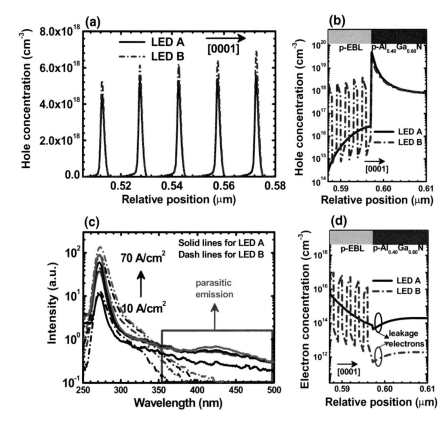

Fig. 4.10 Numerically calculated hole concentration profiles in **a** MQWs for LEDs A and B, **b** p-EBLs and p-Al$_{0.40}$Ga$_{0.60}$N layers for LEDs A and B, **c** experimentally measured EL spectra at different injection current density levels, which are plot in semilog scale for LEDs A and B, **d** numerically calculated electron concentration profiles in p-EBLs and p-Al$_{0.40}$Ga$_{0.60}$N layers for LEDs A and B. Data for Fig. **a**, **b** and **d** are calculated at the current density of 50 A/cm^2. Reproduced from Ref. [13], with the permission of Springer

In summary, the advantage of the specifically designed superlattice p-EBL has been demonstrated. The proposed structure can significantly increase the hole concentration in the superlattice p-EBL, which correspondingly improves the hole injection into the MQW region. The enhanced hole concentration in the MQWs enables the more efficient electron-hole radiative recombination rate, which helps to suppress the electron leakage. In the meanwhile, the specifically designed superlattice p-EBL also increases the electron blocking effect by the last quantum barrier. As a result, the EQE has been improved and the nearly-efficiency-droop performance has been obtained. Although the superlattice p-EBL is super in improving the EQE for DUV LEDs, one shall pay more attention when growing the very thin layer.

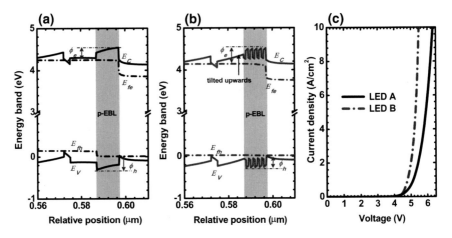

Fig. 4.11 Numerically calculated energy band diagrams in the vicinity of the p-EBLs for **a** LED A and **b** LED B, **c** experimentally measured current as a function of the applied bias. E_C, E_V, $\varnothing_e$ and $\varnothing_h$ denote the conduction band, the valence band, the effective barrier heights for conduction band and valence band, respectively. Fig. **a** and **b** are calculated at the current density of 50 A/cm^2. Reproduced from Ref. [13], with the permission of Springer

4.3 Manipulate the Hole Injection Mechanism by Using Novel p-EBLs Structure for DUV LEDs

It is also unambiguous that the hole injection can be improved by reducing the p-EBL thickness, since by doing so, the intraband tunneling efficiency for holes can be enhanced [22]. The intraband tunneling process can also be enabled for a thick p-EBL according to our previous report [23]. We insert a very thin AlGaN layer in the p-EBL, and the AlN composition for the AlGaN insertion layer is lower than that for layers L1 and L2 [see Fig. 4.12b]. Meanwhile, we purposely make layer L2 thin so that both the thermionic emission (P0) and the intraband tunneling process (P1) can be simultaneously allowed. Then the hole concentration in the thin AlGaN layer will become high which is very useful to reduce the valence band barrier for layer L1, and thus facilitating the thermionic emission of P2. The valence band diagrams for the DUV LEDs are presented in Fig. 4.12c and d, respectively. The calculated barrier heights for the two p-EBLs are shown in Table 4.2. We can see that the valence band barrier height in layer L1 for holes is smaller, which promises the enhanced hole injection efficiency by using the proposed p-EBL architecture in Fig. 4.12b.

To more clearly address the advantage of the p-AlGaN/AlGaN/p-AlGaN EBL over the conventional p-AlGaN EBL in increasing the hole injection efficiency, we calculate and show the hole concentration profiles in the vicinity of the p-EBL and in the MQWs in Fig. 4.13a and b for Devices A and B, respectively. Figure 4.13a demonstrates that the hole accumulation between the p-EBL and the p-Al$_{0.40}$Ga$_{0.60}$N layer decreases if we compare Device B to Device A. The reduced hole accumula-

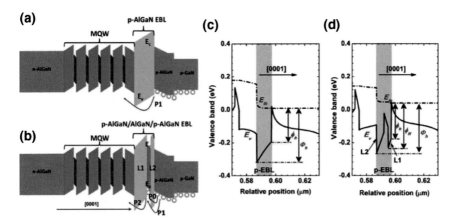

Fig. 4.12 Schematic energy band diagrams for **a** DUV LED with p-AlGaN EBL, **b** DUV LED with p-AlGaN/AlGaN/p-AlGaN EBL, **c** valence band for p-AlGaN EBL, and **d** valence band for p-AlGaN/AlGaN/p-AlGaN EBL. Φ_h, ϕ_H, and ϕ_h denote the effective valence band barrier height at different positions for both p-EBLs. E_v and E_{fh} represent the valence band edge and the quasi-Fermi level for holes, respectively. Reproduced from Ref. [23], with the permission of American Chemistry Society

Table 4.2 Calculated values of Φ_h, ϕ_H, and ϕ_h for Devices A and B

–	Device A	Device B
ϕ_h (meV)	~206.51	~217.40
ϕ_H (meV)	–	~234.64
Φ_h (meV)	~335.18	~303.41

Reproduced from Ref. [23], with the permission of American Chemistry Society

tion level at the p-EBL/p-Al$_{0.40}$Ga$_{0.60}$N interface for Device B reflects the reduced hole blocking effect by the p-AlGaN/AlGaN/p-AlGaN EBL. To further prove the effectiveness of the p-AlGaN/AlGaN/p-AlGaN EBL in enhancing the hole injection capability, the hole concentration profiles within the MQWs for Devices A and B are shown in Fig. 4.13b, which clearly addresses that the hole concentration in the MQWs for Device B is higher than that for Device A. Because of the enhanced hole concentration level, the radiative recombination rate for Device B increases as presented in Fig. 4.13c.

Then, we show the experimentally measured and numerically calculated EQE and optical power in terms of the injection current for Devices A and B in Fig. 4.14a and b, respectively. From the perspective of the measurement and the calculation, the EQE and the optical power are both improved for Device B, e.g., the power is enhanced by ~19.38% at the current of 250 mA according to Fig. 4.14a. On the other hand, the agreement between the experimentally measured results and the numerically calculated results validate the physical models and the parameters that are adopted during our numerical simulations, which further illustrates that the

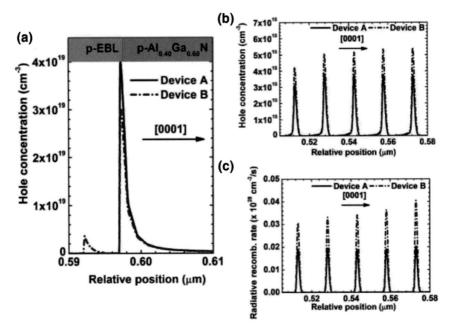

Fig. 4.13 **a** Hole concentration profiles in the p-EBL and the p-Al$_{0.40}$Ga$_{0.60}$N layer, **b** hole concentration profiles in the MQWs, and **c** radiative recombination rate in the MQWs for Devices A and B. Fig. **a**, **b** and **c** are calculated at the current of 100 mA. The chip size is $650 \times 320 \, \mu m^2$. Reproduced from Ref. [23], with the permission of American Chemistry Society

proposed p-AlGaN/AlGaN/p-AlGaN EBL reduces the hole blocking effect by the p-EBL and eventually increases the hole injection efficiency for AlGaN based DUV LEDs.

Furthermore, we also investigate the sensitivity of the hole injection to different p-AlGaN/AlGaN/p-AlGaN EBLs, and our studies show that the thickness, the position and the AlN composition for the AlGaN insertion layer have to be optimized, e.g., a too thick AlGaN insertion layer cannot effectively confine electrons [24]; we also have to keep layer L2 thin and the AlN composition of the AlGaN insertion layer proper for enabling the intraband tunneling process of P0 [25]. In this section, we will highlight the importance of the thickness of the layer L2 of Fig. 4.12b in affecting the hole injection.

When we discuss Fig. 4.12b, we have predicted that layer L2 in Fig. 4.12b has to be made thin so that the intraband tunneling for holes can be significant. To further prove it, we design the devices that are shown in Fig. 4.15. The reference device has the 10 nm thick p-Al$_{0.60}$Ga$_{0.40}$N as the EBL. The proposed devices utilize the 10 nm thick p-Al$_{0.60}$Ga$_{0.40}$N(L$_1$)/p-Al$_y$Ga$_{1-y}$N/p-Al$_{0.60}$Ga$_{0.40}$N(L$_2$) EBLs, for which the AlN composition for the 7 nm p-Al$_y$Ga$_{1-y}$N layer and the thickness for the p-Al$_{0.60}$Ga$_{0.40}$N layers are variables.

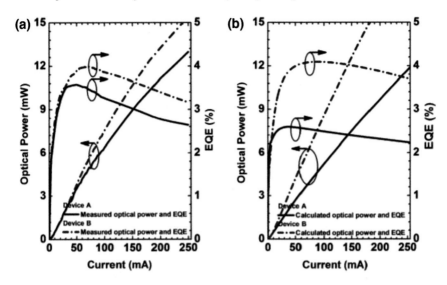

Fig. 4.14 a Experimentally measured and **b** numerically calculated EQE and optical power for Devices A and B. Reproduced from Ref. [23], with the permission of American Chemistry Society

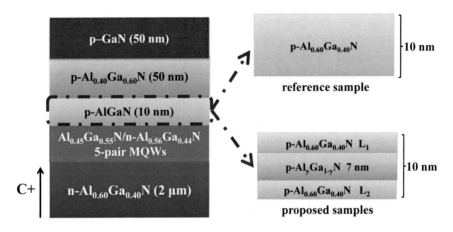

Fig. 4.15 Schematic device architectures for the reference device with the 10 nm thick p-$Al_{0.60}Ga_{0.40}N$ as the EBL and the proposed devices with the 10 nm thick p-$Al_{0.60}Ga_{0.40}N(L_1)$/p-$Al_yGa_{1-y}N$/p-$Al_{0.60}Ga_{0.40}N$ (L_2) EBLs. Reproduced from Ref. [25], with the permission of SPIE

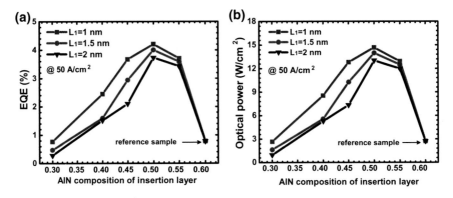

Fig. 4.16 Numerically calculated **a** EQE and **b** optical power density for the investigated DUV LEDs. Reproduced from Ref. [25], with the permission of SPIE

To probe the impact of the AlN composition in the p-$Al_yGa_{1-y}N$ layer and the thickness for the p-$Al_{0.60}Ga_{0.40}N$ layers on DUV LEDs, we present the EQE and the optical power density in Fig. 4.16a and b, respectively. Clearly we see that the EQE and the optical power density increase as the AlN composition in the p-$Al_yGa_{1-y}N$ insertion layer increases to 50%. When the AlN composition in the p-$Al_yGa_{1-y}N$ insertion layer is beyond 50%, then the EQE and the optical power density decreases.

The different optical performances in Fig. 4.16a and b arise from the hole injection, which can be explained by using the energy band diagrams in Fig. 4.17a, b and c. We selectively show the devices with the AlN compositions of 0.45, 0.50 and 0.55 for the p-$Al_yGa_{1-y}N$ insertion layers, and the layer L_1 thickness is kept to be 1 nm. According to the energy bands, we can get that the hole injection is co-determined by Φ_b, ϕ_B and ϕ_b, which denote the effective valence band barrier heights at different positions for the p-EBL, respectively. $\Delta\Phi$ reflects the tilted level of the energy band for the last quantum well. The holes are injected into the p-$Al_yGa_{1-y}N$ insertion layer, which process is realized by thermally assisted intraband tunneling, such that the holes tunnel through layer L_1 and reach the p-$Al_yGa_{1-y}N$ insertion layer by climbing over the barrier height of ϕ_B. Hence the value of ϕ_B strongly affects the hole injection. However ϕ_B is decided by the AlN composition in the p-$Al_yGa_{1-y}N$ insertion layer. On the other hand, the hole transport is also subject to the barrier height of ϕ_b, the value for which influences the thermionic emission. The value for ϕ_b can be increased by making layer L_1 thick. Once the holes are stored in the p-$Al_yGa_{1-y}N$ insertion layer, their injection into the MQWs is strongly affected by Φ_b and $\Delta\Phi$. A small Φ_b and a less tilted valence band for the last quantum barrier (LQB) are preferable for the improved hole injection.

Table 4.3 summarizes the values of Φ_b, ϕ_B, ϕ_b and $\Delta\Phi$ for the investigated p-EBLs. When the AlN composition in the p-$Al_yGa_{1-y}N$ insertion layer increases, the ϕ_B simultaneously becomes large, which hinders the hole injection from the p-type hole injector into the p-EBL [see Fig. 4.17d]. Nevertheless, the increase of the AlN composition for the p-$Al_yGa_{1-y}N$ insertion layer can reduce the valence band

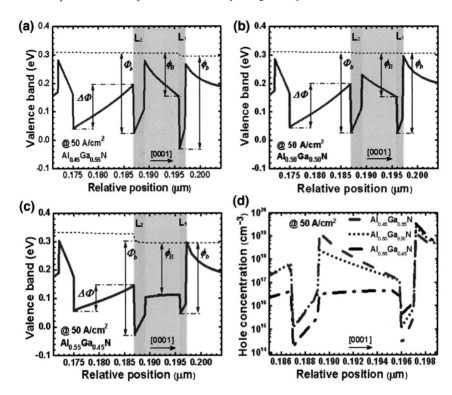

Fig. 4.17 Valence band diagrams for the p-EBLs with **a** Al$_{0.45}$Ga$_{0.55}$N, **b** Al$_{0.50}$Ga$_{0.50}$N, and **c** Al$_{0.55}$Ga$_{0.45}$N insertion layers, and **d** hole concentration profiles in the p-EBLs. Φ_b, ϕ_B and ϕ_b denote the effective valence band barrier heights at different positions for the p-EBL. $\Delta\Phi$ reflects the tilted level of the energy band for the LQB. Data are calculated at the injection current level of 50 A/cm^2. Reproduced from Ref. [25], with the permission of SPIE

Table 4.3 Calculated values for ϕ_b, ϕ_B, Φ_b, and $\Delta\Phi$ when the injection current density is 50 A/cm^2

AlN Composition for Al$_y$Ga$_{1-y}$N	ϕ_b (meV)	ϕ_B (meV)	Φ_b (meV)	$\Delta\Phi$ (meV)
y = 0.45	325.7	151.8	281.2	152.1
y = 0.50	278.1	155.0	278.7	149.9
y = 0.55	245.6	183.3	329.2	87.2

Reproduced from Ref. [25], with the permission of SPIE

offset between the p-Al$_y$Ga$_{1-y}$N insertion layer and layer L$_2$, which correspondingly decreases Φ_b. Hence, the comprised design for the best hole injection shall be the one with p-Al$_{0.50}$Ga$_{0.50}$N insertion layer. Further investigations into Fig. 4.17a–d and Table 4.3 also imply that the thermionic emission for holes is not significant when the holes are transporting through layer L1 if layer L1 is 1 nm thick.

Figure 4.16a and b also indicate that the EQE and the optical power density increase with the decreasing thickness for layer L$_1$. This conclusion is consistent

Table 4.4 Values of ϕ_b, Φ_b and $\Delta\Phi$ for the p-EBL with the Al$_{0.50}$Ga$_{0.50}$N insertion layer near the p-region (L$_1$ = 1 nm), at middle (L$_1$ = 1.5 nm), and near the MQWs (L$_1$ = 2 nm) at the injection current density of 50 A/cm^2

Position for Al$_{0.50}$Ga$_{0.50}$N insertion layer (nm)	ϕ_b (meV)	Φ_b (meV)	$\Delta\Phi$ (meV)
L$_1$ = 1	278.1	278.7	149.9
L$_1$ = 1.5	296.1	272.3	159.6
L$_1$ = 2	314.0	266.6	165.4

Reproduced from Ref. [25], with the permission of SPIE

with our prediction in Fig. 4.12b. Here, to clarify the origin for the observations in Fig. 4.16a and b, we selectively investigate the barrier heights for the DUV LEDs with p-Al$_{0.60}$Ga$_{0.40}$N(L$_1$)/p-Al$_{0.50}$Ga$_{0.50}$N/p-Al$_{0.60}$Ga$_{0.40}$N (L$_2$) EBLs, for which the values of L$_1$ are set to 1 nm, 1.5 nm and 2 nm, respectively. The definition of the symbols for the barrier height is identical to the one in Fig. 4.17a–c. We only summarized the values in Table 4.4, which shows that, when the layer L$_1$ becomes thin, the value of ϕ_b gets small. A small ϕ_b means that the thermionic emission rate is increased when the holes are injected through layer L$_1$. Meanwhile, a thin layer L$_1$ also facilitates the intraband tunneling for holes, which helps to increase the hole concentration in the p-Al$_{0.50}$Ga$_{0.50}$N insertion layer. The high hole concentration in the p-Al$_{0.50}$Ga$_{0.50}$N insertion layer helps to reduce the valence band barrier in layer L$_2$ for the holes, and thus the value of Φ_b decreases with the reducing thickness for layer L$_1$ according to Table 4.4.

Therefore, we have demonstrated to manipulate the hole injection by using the p-AlGaN/AlGaN/p-AlGaN EBL, which can maintain both the high-efficiency intraband tunneling and the excellent thermionic emission for holes. Nevertheless, we also make the parametric investigation regarding different p-AlGaN/AlGaN/p-AlGaN EBLs. We find that the AlN composition and the thickness for the AlGaN insertion layer have to be optimized. Otherwise, the electron leakage can be severe. In addition, the AlGaN insertion layer has to be close to the p-type hole injection layer. By doing so, the p-AlGaN layer that is close to the p-type hole injection layer can be thin, which then can favor both the intraband tunneling and the thermionic emission for the holes. Once the AlGaN insertion layer is thin, it can readily possess a very high hole concentration. The high hole concentration in the AlGaN thin layer enables the reduced hole blocking effect that is caused by the p-AlGaN layer (the p-AlGaN layer here is the one close to the LQB). Then, the hole injection into the MQW region can be enhanced simultaneously.

4.4 Increase the Hole Concentration in the MQWs for DUV LEDs

The hole injection is strongly hindered by the quantum barriers and the holes are often accumulated in the last quantum well that is closest to the p-EBL for InGaN/GaN based visible LEDs [26]. However, the hole transport within the MQWs for DUV

LEDs shows different concentration profiles according to the reports by different groups [13, 14, 18, 23, 27–31], such that the highest hole concentration is not always found in the last quantum well closest to the p-EBL [14, 28, 30, 31], while the hole accumulation takes place in the last quantum well in other reports [13, 18, 23, 27]. These observations are attributed to the reduced valence band offset for the $Al_xGa_{1-x}N/Al_yGa_{1-y}N$ based quantum wells if different x and y values are used [28]. In addition, different numerical models have assumed various energy band offset ratio between the conduction band offset and the valence band offset, which also remarkably affect the hole transport within the active region. Some typical energy band offset ratios include 70/30 [27, 29, 30, 32], 65/35 [14, 28] and 50/50 [13, 18, 23]. Interestingly, our studies also show that, when compared to InGaN/GaN based visible LEDs, the hole distribution across the active region is more uniform despite the 50/50 band offset ratio [13, 18, 23]. Therefore, we suggest that efforts shall be made to increase the hole concentration level in each quantum well rather than homogenizing the hole profiles across the active region for DUV LEDs [20].

References

1. Katsuragawa M, Sota S, Komori M, Anbe C, Takeuchi T, Sakai H, Amano H, Akasaki I (1998) Thermal ionization energy of Si and Mg in AlGaN. J Cryst Growth 189:528–531. https://doi.org/10.1016/S0022-0248(98)00345-5
2. Zhang L, Ding K, Yan JC, Wang JX, Zeng YP, Wei TB, Li YY, Sun BJ, Duan RF, Li JM (2010) Three-dimensional hole gas induced by polarization in (0001)-oriented metal-face III-nitride structure. Appl Phys Lett 97(6):062103. https://doi.org/10.1063/1.3478556
3. Simon J, Protasenko V, Lian C, Xing H, Jena D (2010) Polarization-induced hole doping in wide-band-gap uniaxial semiconductor heterostructures. Science 327(5961):60–64. https://doi.org/10.1126/science.1183226
4. Bayram C, Pau JL, McClintock R, Razeghi M (2008) Delta-doping optimization for high quality p-type GaN. J Appl Phys 104(8):083512. https://doi.org/10.1063/1.3000564
5. Bayram C, Pau JL, McClintock R, Razeghi M (2008) Performance enhancement of GaN ultraviolet avalanche photodiodes with p-type δ-doping. Appl Phys Lett 92(24):241103. https://doi.org/10.1063/1.2948857
6. Li T, Simbrunner C, Wegscheider M, Navarro-Quezada A, Quast M, Schmidegg K, Bonanni A (2008) GaN:δ-Mg grown by MOVPE: structural properties and their effect on the electronic and optical behavior. J Cryst Growth 310(1):13–21. https://doi.org/10.1016/j.jcrysgro.2007.09.045
7. Gunning B, Lowder J, Moseley M, Doolittle WA (2012) Negligible carrier freeze-out facilitated by impurity band conduction in highly p-type GaN. Appl Phys Lett 101(8):082106. https://doi.org/10.1063/1.4747466
8. Chen Y, Wu H, Han E, Yue G, Chen Z, Wu Z, Wang G, Jiang H (2015) High hole concentration in p-type AlGaN by indium-surfactant-assisted Mg-delta doping. Appl Phys Lett 106(16):162102. https://doi.org/10.1063/1.4919005
9. Kim JK, Waldron EL, Li Y-L, Gessmann T, Schubert EF, Jang HW, Lee J-L (2004) P-type conductivity in bulk $Al_xGa_{1-x}N$ and $Al_xGa_{1-x}N/Al_yGa_{1-y}N$ superlattices with average Al mole fraction > 20%. Appl Phys Lett 84(17):3310–3312. https://doi.org/10.1063/1.1728322
10. Cheng B, Choi S, Northrup JE, Yang Z, Knollenberg C, Teepe M, Wunderer T, Chua CL, Johnson NM (2013) Enhanced vertical and lateral hole transport in high aluminum-containing

AlGaN for deep ultraviolet light emitters. Appl Phys Lett 102(23):231106. https://doi.org/10. 1063/1.4809947

11. Kipshidze G, Kuryatkov V, Zhu K, Borisov B, Holtz M, Nikishin S, Temkin H (2003) AlN/AlGaInN superlattice light-emitting diodes at 280 nm. J Appl Phys 93(3):1363–1366. https://doi.org/10.1063/1.1535255

12. Sergey N, Vladimir VK, Anilkumar C, Boris AB, Gela DK, Iftikhor A, Mark H, Henryk T (2003) Deep ultraviolet light emitting diodes based on short period superlattices of AlN/AlGa(In)N. Jpn J Appl Phys 42(11B):L1362. https://doi.org/10.1143/JJAP.42.L1362

13. Zhang Z-H, Chen S-WH, Chu C, Tian K, Fang M, Zhang Y, Bi W, Kuo H-C (2018) Nearly efficiency-droop-free AlGaN-based ultraviolet light-emitting diodes with a specifically designed superlattice p-type electron blocking layer for high Mg doping efficiency. Nanoscale Res Lett 13:122. https://doi.org/10.1186/s11671-018-2539-9

14. Kuo Y-K, Chang J-Y, Chen F-M, Shih Y-H, Chang H-T (2016) Numerical Investigation on the carrier transport characteristics of AlGaN deep-UV light-emitting diodes. IEEE J Quantum Electron 52(4):1–5. https://doi.org/10.1109/JQE.2016.2535252

15. Kuo Y, Chang J, Chang H, Chen F, Shih Y, Liou B (2017) Polarization effect in AlGaN-based deep-ultraviolet light-emitting diodes. IEEE J Quantum Electron 53(1):1–6. https://doi.org/10. 1109/JQE.2016.2643289

16. Zhang Z-H, Liu W, Tan ST, Ji Y, Wang L, Zhu B, Zhang Y, Lu S, Zhang X, Hasanov N, Sun XW, Demir HV (2014) A hole accelerator for InGaN/GaN light-emitting diodes. Appl Phys Lett 105(15):153503. https://doi.org/10.1063/1.4898588

17. Zhang Z-H, Zhang Y, Bi W, Geng C, Xu S, Demir HV, Sun XW (2016) On the hole accelerator for III-nitride light-emitting diodes. Appl Phys Lett 108(15):071101. https://doi.org/10.1063/ 1.4947025

18. Zhang Z-H, Li L, Zhang Y, Xu F, Shi Q, Shen B, Bi W (2017) On the electric-field reservoir for III-nitride based deep ultraviolet light-emitting diodes. Opt Express 25(14):16550–16559. https://doi.org/10.1364/OE.25.016550

19. Zhang Z-H, Liu W, Ju Z, Tan ST, Ji Y, Zhang X, Wang L, Kyaw Z, Sun XW, Demir HV (2014) Polarization self-screening in [0001] oriented InGaN/GaN light-emitting diodes for improving the electron injection efficiency. Appl Phys Lett 104(25):251108. https://doi.org/10.1063/1. 4885421

20. Zhang Z-H, Chu C, Chiu CH, Lu TC, Li L, Zhang Y, Tian K, Fang M, Sun Q, Kuo H-C, Bi W (2017) UVA light-emitting diode grown on Si substrate with enhanced electron and hole injections. Opt Lett 42(21):4533–4536. https://doi.org/10.1364/OL.42.004533

21. Kolbe T, Stellmach J, Mehnke F, Rothe MA, Kueller V, Knauer A, Einfeldt S, Wernicke T, Weyers M, Kneissl M (2016) Efficient carrier-injection and electron-confinement in UV-B light-emitting diodes. Phys Status Solidi a-Appl Mater Sci 213(1):210–214. https://doi.org/10. 1002/pssa.201532479

22. Mehnke F, Kuhn C, Guttmann M, Reich C, Kolbe T, Kueller V, Knauer A, Lapeyrade M, Einfeldt S, Rass J, Wernicke T, Weyers M, Kneissl M (2014) Efficient charge carrier injection into sub-250 nm AlGaN multiple quantum well light emitting diodes. Appl Phys Lett 105(5):051113. https://doi.org/10.1063/1.4892883

23. Zhang Z-H, Chen S-WH, Zhang Y, Li L, Wang S-W, Tian K, Chu C, Fang M, Kuo H-C, Bi W (2017) Hole transport manipulation to improve the hole injection for deep ultraviolet light-emitting diodes. Acs Photonics 4(7):1846–1850. https://doi.org/10.1021/acsphotonics. 7b00443

24. Chu CS, Tian KK, Fang MQ, Zhang YH, Li LP, Bi WG, Zhang ZH (2018) On the $Al_xGa_{1-x}N/Al_yGa_{1-y}N/Al_xGa_{1-x}N$ (x > y) p-electron blocking layer to improve the hole injection for AlGaN based deep ultraviolet light-emitting diodes. Superlattices Microstruct 113:472–477. https://doi.org/10.1016/j.spmi.2017.11.029

25. Chu C, TianK, FangM, ZhangY, Zhao S, Bi W, Zhang Z-H (2018) Structural design and optimization of deep-ultraviolet light-emitting diodes with $Al_xGa_{1-x}N/Al_yGa_{1-y}N/AlxGa_{1-x}N$ (x > y) p-electron blocking layer. J Nanophotonics 12(4):043503, May 2018. https://doi.org/ 10.1117/1.jnp.12.043503

26. Zhang Z-H, Zhang Y, Bi W, Demir HV, Sun XW (2016) On the internal quantum efficiency for InGaN/GaN light-emitting diodes grown on insulating substrates. Phys Status Solidi (a) 213(12):3078–3102. https://doi.org/10.1002/pssa.201600281
27. Zhang M, Li Y, Chen S, Tian W, Xu J, Li X, Wu Z, Fang Y, Dai J, Chen C (2014) Performance improvement of AlGaN-based deep ultraviolet light-emitting diodes by using staggered quantum wells. Superlattices Microstruct 75:63–71. https://doi.org/10.1016/j.spmi.2014.07.002
28. Tsai M-C, Yen S-H, Kuo Y-K (2011) Deep-ultraviolet light-emitting diodes with gradually increased barrier thicknesses from n-layers to p-layers. Appl Phys Lett 98(11):111114. https://doi.org/10.1063/1.3567786
29. Yang GF, Xie F, Dong KX, Chen P, Xue JJ, Zhi T, Tao T, Liu B, Xie ZL, Xiu XQ, Han P, Shi Y, Zhang R, Zheng YD (2014) Design of deep ultraviolet light-emitting diodes with staggered AlGaN quantum wells. Physica E 62:55–58. https://doi.org/10.1016/j.physe.2014.04.014
30. Yin YA, Wang N, Fan G, Zhang Y (2014) Investigation of AlGaN-based deep-ultraviolet light-emitting diodes with composition-varying AlGaN multilayer barriers. Superlattices Microstruct 76:149–155. https://doi.org/10.1016/j.spmi.2014.10.003
31. Tian K, Chen Q, Chu C, Fang M, Li L, Zhang Y, Bi W, Chen C, Zhang Z-H, Dai J (2018) Investigations on AlGaN-based deep-ultraviolet light-emitting diodes with Si-doped quantum barriers of different doping concentrations. Physica Status Solidi-Rapid Res Lett 12(1):1700346. https://doi.org/10.1002/pssr.201700346
32. Kim SJ, Kim TG (2014) Deep-ultraviolet AlGaN light-emitting diodes with variable quantum well and barrier widths. Physica Status Solidi (a) 211(3):656–660. https://doi.org/10.1002/pssa.201330258

Chapter 5
Enhance the Electron Injection Efficiency for DUV LEDs

Abstract The unbalanced carrier injection for DUV LEDs illustrates that the electron tends to overflow from the active region. The underly mechanism arises from three aspects: (1) electrons cannot be consumed by forming electron-hole pairs and recombine radiatively in the active region, which is due to the insufficient hole injection, (2) the electron have larger mobility and are more mobile, (3) The conduction band offset between the AlGaN based quantum barrier and quantum well decreases, which correspondingly reduces the conduction band barrier height, enabling the active region to lose the effective confinement capability for electrons. In this chapter, we propose different methods for increasing the electron injection efficiency, and specifically, we demonstrate novel designs to reduce the electron drift velocity and hence the energy, so that the quantum wells have more chances of capturing the electrons.

Besides engineering the hole injection, the electron injection shall also be manipulated, for which extensive research efforts have been made for InGaN/GaN based visible LEDs [1, 2]. However, AlGaN based DUV LEDs are also influenced by the unbalanced carrier injection, such that the electron injection has to be enhanced [3]. A direct method to suppress the electron leakage for DUV LEDs is to engineer the p-EBL, e.g., superlattice p-EBL [4–7], p-EBL with AlGaN insertion layer [8–10], p-EBL with the graded AlN composition [11, 12], superlattice last quantum barrier [13]. Moreover, the electron concentration in the MQWs can be enhanced by proposing novel active region structure, e.g., Si doped quantum barriers [14], grading the AlN composition for the AlGaN based quantum barriers [15], increasing the AlN composition for AlGaN based quantum barrier [16].

Most recently, our group has reported to manipulate the electron drift velocity and the electron energy for DUV LEDs, and the electron injection efficiency can be enhanced [17, 18]. The electron energy can be tuned by modulating the Si doping concentration for the n-AlGaN layer, and by doing so, the n-AlGaN/first quantum barrier interface can generate the build-in electric field, which can reduce the electron drift velocity and the electron energy [18]. Once the electrons are "cooled down", the MQWs can have more chances to capture the electrons. Besides using the modulated Si doping concentration in the n-AlGaN layer, the electrons can also be "cooled

© The Author(s), under exclusive license to Springer Nature Singapore Pte Ltd. 2019   33
Z.-H. Zhang et al., *Deep Ultraviolet LEDs*, Nanoscience and Nanotechnology,
https://doi.org/10.1007/978-981-13-6179-1_5

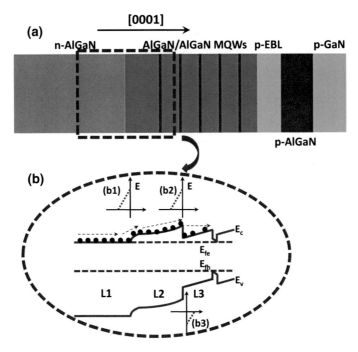

Fig. 5.1 a Schematic structure for the [0001] oriented DUV LEDs, **b** schematic energy band diagram when the electron concentration and the alloy in the n-AlGaN layer are modulated (i.e., the electron concentration for the L2 region is lower than that for the L1 region; the AlN composition for the L3 region is lower than that for the L2 region). The positive direction of the electric field is along the [0001] orientation. The L1/L2 interface possesses the electric field along the [0001] orientation as shown in Fig. (b1). The sketched electric field profiles at the L2/L3 interface are presented in Fig. (b2) and (b3), i.e., the electric fields on the L2 side and on the L3 side are along and opposed to the [0001] orientation, respectively. E_c, E_v, E_{fe} and E_{fh} represent the conduction band, the valence band, quasi-Fermi levels for electrons and holes, respectively. Reproduced from Ref. [17], with the permission of Optical Society of America

down" by using the polarization induced electric field [17]. The underlying device physics is depicted in Fig. 5.1. The schematic device architecture for the studied [0001] oriented DUV LEDs is shown in Fig. 5.1a. On one hand, the different doping concentration levels within the n-AlGaN layer can cause an interface depletion region, and the electric field in the depletion region is along the [0001] orientation [see Fig. 5.1b1, in which the electron concentration for the L2 region is lower than that for the L1 region], which can reduce the kinetic energy for electrons, and on the other hand, the $Al_x G_{1-x}N/Al_y Ga_{1-y}N$ (x > y) interface possesses the negative polarization induced interface charges that can simultaneously produce the polarization induced electric field. More importantly, the polarization induced electric field on the $Al_x G_{1-x}N$ side is along the [0001] orientation [e.g., the L2/L3 interface in Fig. 5.1b2] that helps to reduce the kinetic energy for the incoming free electrons. It shall be noted that the electric field on the $Al_y G_{1-y}N$ side is opposite to the [0001] orientation [e.g., the L2/L3 interface in Fig. 5.1b3], and the electrons will obtain more energy when traveling through it. Fortunately, the electric field intensity in Fig. 5.1b3 is smaller than that in Fig. 5.1b2 since the high AlN composition in the $Al_xG_{1-x}N$ layer causes a smaller dielectric constant, and this enables the even stronger electric field intensity therein [19, 20]. The structural information for different n-AlGaN electron injectors is illustrated in Table 5.1, which includes seven device structures with various AlN compositions and Si doping concentrations in the n-AlGaN layers.

Figure 5.2a and b show the numerically calculated and the experimentally measured EQE, optical power density and the current density-voltage characteristics, respectively for Device 1. The experimental EQE and optical power density show excellent agreement with the calculated ones. Meanwhile, the measured current density in terms of the applied voltage is also reproduced numerically as illustrated in Fig. 5.2b. Figure 5.2a and b indicate the reasonable physical models and parameters are set in our calculations.

To support our speculations in Fig. 5.1a and b, we selectively show the energy band diagrams for Devices 1, 2, 3 and 7 in Fig. 5.3a–e, respectively. Figure 5.3a and b present the energy band for Device 1 with zoom-in scale in Fig. 5.3b. The electron injection layer is purely n-$Al_{0.58}Ga_{0.42}N$ layer with the constant Si doping concentration of 3×10^{18} cm^{-3}. The energy band is flat and the electric field in the n-$Al_{0.58}Ga_{0.42}N$ layer is small according to Fig. 5.3a. Moreover, we utilize the undoped $Al_{0.55}Ga_{0.45}N$ as the quantum barrier material. Therefore, the negative polarization induced sheet charges can be generated for the n-$Al_{0.58}Ga_{0.42}N$ layer/$Al_{0.55}Ga_{0.45}N$ layer interface, which can bend the energy band upwards and produce very strong

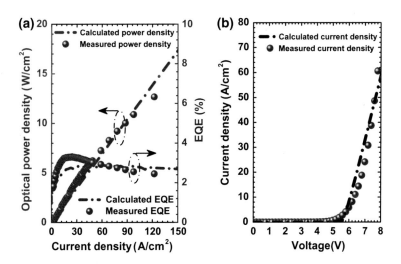

Fig. 5.2 a Numerically calculated and experimentally measured EQE and optical power density as a function of the current density for Device 1, **b** numerically calculated and experimentally measured current density in terms of the applied voltage for Device 1. Reproduced from Ref. [17], with the permission of Optical Society of America

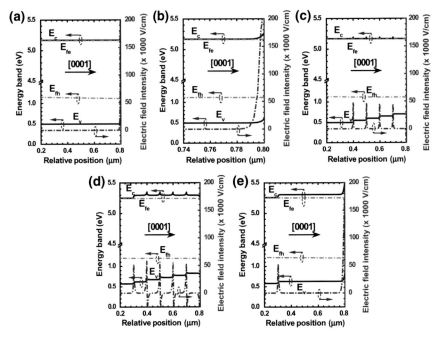

Fig. 5.3 **a** Energy band and electric field profiles in the n-AlGaN layers for (**a**) Device 1, **b** Device 1 with zoom-in range, **c** Device 2, **d** Device 3 and **e** Device 7. All figures are calculated at the current density of 160 A/cm^2. E_c, E_v, E_{fe} and E_{fh} denote the conduction band, the valence band, the quasi-Fermi levels for electrons and holes, respectively. Reproduced from Ref. [17], with the permission of Optical Society of America

electric field intensity at the n-Al$_{0.58}$Ga$_{0.42}$N/Al$_{0.55}$Ga$_{0.45}$N interface, i.e., interface depletion. The interface depletion region width in the n-Al$_{0.58}$Ga$_{0.42}$N layer is narrow, which can be observed by zooming in the scale as shown in Fig. 5.3b. Figure 5.3c presents the energy band and the electric field profile for Device 2. As has been illustrated in Table 5.1, the n-AlGaN electron injection layer possesses the stair-cased AlN composition such that the AlN composition decreases along the [0001] orientation. The stair-cased decrease for the AlN composition of the n-AlGaN layer is also indicated by Fig. 5.3c. The stair-cased AlN composition for the n-AlGaN electron injection layer enables the generation of the very strong polarization induced electric field as shown in Fig. 5.3c. The electric field is along the [0001] orientation, which helps to decelerate the electrons and make them less "hot". The calculated energy band diagram and the electric field profile for Device 3 are presented in Fig. 5.3d. Device 3 differs from Deice 2 only in the Si doping concentration for the n-Al$_{0.56}$Ga$_{0.44}$N layer, n-Al$_{0.54}$Ga$_{0.46}$N layer, n-Al$_{0.52}$Ga$_{0.48}$N layer and n-Al$_{0.50}$Ga$_{0.50}$N layer, which is 3×10^{17} cm^{-3} rather than 3×10^{18} cm^{-3}. According to Fig. 5.3d, the electric field that is along the [0001] orientation occurs at the Al$_y$Ga$_{1-y}$N side of the Al$_x$G$_{1-x}$N/Al$_y$Ga$_{1-y}$N ($x > y$) interface, e.g., the n-Al$_{0.54}$Ga$_{0.46}$N side of the n-$Al_{0.56}Ga_{0.44}N$/n-$Al_{0.54}Ga_{0.46}N$ interface, the n-$Al_{0.52}Ga_{0.48}N$ side of the n-$Al_{0.54}Ga_{0.46}N$/n-$Al_{0.52}Ga_{0.48}N$ interface. The electric field intensity is lower than the counterpart, which is consistent with our predictions in Fig. 5.1b3, such that the larger dielectric constant for the $Al_yGa_{1-y}N$ side reduces the electric field intensity. In addition, the conduction band barrier height is observed at the n-$Al_{0.58}Ga_{0.42}N$/n-$Al_{0.56}Ga_{0.44}N$ interface. However, this barrier height is absent in Fig. 5.3c. We believe this is caused by the interface depletion due to the modulated Si doping concentration. Figure 5.3e contains the energy band and the electric field profile for Device 7. The n-AlGaN electron injection layer for Device 7 has the identical AlN composition except that the Si doping concentration is modulated according to Table 5.1. The interface depletion region can produce the electric field as shown in Fig. 5.3e. Details of the generation mechanism for the electric field when the Si doping concentration is modulated have been discussed in our previous report [18].

After calculating the electric field profiles, by using $W = e \int_0^l E_{field} \cdot dx$ (e, l and E_{field} represent the electronic unit charge and the electric field profiles within the integration range of l, respectively. dx is the integration step that has been properly optimized when setting the mesh lines during numerical computations), we can calculate the net work (ΔE) that is conducted on the electrons during the transportation in the AlGaN layers for Device 1–7. For comparative study, the l ranges from 0.2 to 0.8 μm which has been shown in Fig. 5.3a–e. In our case, the electrons will lose energy if the integral is a positive value. The energy that the electrons lose for different devices is summarized in Fig. 5.4a, and Fig. 5.4a also shows that the optical power improves as the electrons are more decelerated. Further insight into Fig. 5.4b tells that once the Si doping concentration in the n-AlGaN layer is too low, the forward voltage can be increased, and this gives rise to a lower WPE. In our work, the thickness for the n-AlGaN layer with the electron concentration of 3×10^{17} cm^{-3} is too thick (i.e., 0.1–0.5 μm). Nevertheless, we believe that the WPE can also be improved once the modulated doped region becomes thin.

The direct impact of the electron energy is on the electron injection efficiency. We then selectively show the electron concentration profiles in the n-AlGaN electron injection layers and the MQW regions for Devices 1 and 5 in Fig. 5.5a and b, respectively. Figure 5.5a shows that the electron concentration significantly decreases in the n-$Al_{0.56}Ga_{0.44}N$, the n-$Al_{0.54}Ga_{0.42}N$ and the n-$Al_{0.52}Ga_{0.46}N$ layers, i.e., the "dips" in the electron concentration profile for Device 5, is attributed to the electron depletion effect by the polarization induced negative changes at the $Al_xGa_{1-x}N$/$Al_yGa_{1-y}N$ ($x > y$) interfaces (i.e., n-$Al_{0.56}Ga_{0.34}N$/n-$Al_{0.54}Ga_{0.36}N$ and n-$Al_{0.54}Ga_{0.36}N$/n-$Al_{0.52}Ga_{0.38}N$ interfaces for Device 5). When we look into Fig. 5.5b, we can see that, except in the first quantum well (the one closest to the n-AlGaN electron injection layer), the electron concentration levels in other four quantum wells for Device 5 have been promoted when compared to that for Device 1. The enhanced electron concentration in the active region for Device 5 is well attributed to the fact that the electrons are less "hot" after being decelerated by the proposed design. The lower Si doping concentration in the n-$Al_{0.56}Ga_{0.34}N$/n-$Al_{0.54}Ga_{0.36}N$/n-$Al_{0.52}Ga_{0.38}N$ layer for Device 5 suppresses the electron diffusion

Table 5.1 Structure information of the thickness, the electron concentration and the alloy composition of the n-AlGaN layers for Devices 1–7. The AlN composition along the [0001] orientation stepwisely decreases for Devices 2, 3, 4, 5 and 6. Reproduced from Ref. [17], with the permission of Optical Society of America

Device number (Di)	Structure information for the respective n-AlGaN layer	
Device 1	n-Al$_{0.58}$Ga$_{0.42}$N (3.0 μm, n = 3 × 10^{18} cm^{-3})	
Device 2	Al$_x$Ga$_{1-x}$N/Al$_y$Ga$_{1-y}$N junctions (x > y)	n-Al$_{0.58}$Ga$_{0.42}$N (2.6 μm, n = 3 × 10^{18} cm^{-3})
		n-Al$_{0.56}$Ga$_{0.44}$N (0.1 μm, n = 3 × 10^{18} cm^{-3})
		n-Al$_{0.54}$Ga$_{0.46}$N (0.1 μm, n = 3 × 10^{18} cm^{-3})
		n-Al$_{0.52}$Ga$_{0.48}$N (0.1 μm, n = 3 × 10^{18} cm^{-3})
		n-Al$_{0.50}$Ga$_{0.50}$N (0.1 μm, n = 3 × 10^{18} cm^{-3})
Device 3	Al$_x$Ga$_{1-x}$N/Al$_y$Ga$_{1-y}$N junctions (x > y)	n-Al$_{0.58}$Ga$_{0.42}$N (2.6 μm, n = 3 × 10^{18} cm^{-3})
		n-Al$_{0.56}$Ga$_{0.44}$N (0.1 μm, n = 3 × 10^{17} cm^{-3})
		n-Al$_{0.54}$Ga$_{0.46}$N (0.1 μm, n = 3 × 10^{17} cm^{-3})
		n-Al$_{0.52}$Ga$_{0.48}$N (0.1 μm, n = 3 × 10^{17} cm^{-3})
		n-Al$_{0.50}$Ga$_{0.50}$N (0.1 μm, n = 3 × 10^{17} cm^{-3})
Device 4	Al$_x$Ga$_{1-x}$N/Al$_y$Ga$_{1-y}$N junctions (x > y)	n-Al$_{0.58}$Ga$_{0.42}$N (2.7 μm, n = 3 × 10^{18} cm^{-3})
		n-Al$_{0.56}$Ga$_{0.44}$N (0.1 μm, n = 3 × 10^{17} cm^{-3})
		n-Al$_{0.54}$Ga$_{0.46}$N (0.1 μm, n = 3 × 10^{17} cm^{-3})
		n-Al$_{0.52}$Ga$_{0.48}$N (0.1 μm, n = 3 × 10^{17} cm^{-3})
Device 5	Al$_x$Ga$_{1-x}$N/Al$_y$Ga$_{1-y}$N junctions (x > y)	n-Al$_{0.58}$Ga$_{0.42}$N (2.8 μm, n = 3 × 10^{18} cm^{-3})
		n-Al$_{0.56}$Ga$_{0.44}$N (0.1 μm, n = 3 × 10^{17} cm^{-3})
		n-Al$_{0.54}$Ga$_{0.46}$N (0.1 μm, n = 3 × 10^{17} cm^{-3})
Device 6	Al$_x$Ga$_{1-x}$N/Al$_y$Ga$_{1-y}$N junction (x > y)	n-Al$_{0.58}$Ga$_{0.42}$N (2.9 μm, n = 3 × 10^{18} cm^{-3})
		n-Al$_{0.56}$Ga$_{0.44}$N (0.1 μm, n = 3 × 10^{17} cm^{-3})
Device 7	n-Al$_{0.58}$Ga$_{0.42}$N (2.5 μm, n = 3 × 10^{18} cm^{-3})	
	n-Al$_{0.58}$Ga$_{0.42}$N (0.5 μm, n = 3 × 10^{17} cm^{-3})	

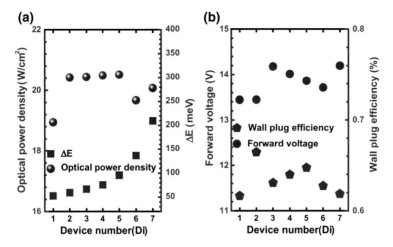

Fig. 5.4 **a** Optical power density and the energy loss of electrons for Devices 1–7, **b** wall-plug efficiency and forward voltage for Devices 1–7. Reproduced from Ref. [17], with the permission of Optical Society of America

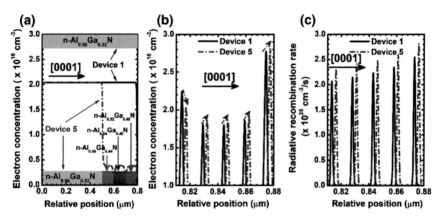

Fig. 5.5 **a** Electron concentration profiles in the n-AlGaN layers, **b** electron concentration profiles in the MQWs, and **c** radiative recombination rates in the MQWs for Devices 1 and 5, respectively. Figures are calculated at the current density of 160 A/cm^2. Reproduced from Ref. [17], with the permission of Optical Society of America

into the first quantum well, and as a result, Device 5 has a lower electron concentration level in the first quantum well than Device 1. Because of the enhanced electron concentration in the last four quantum wells, the radiative recombination is improved for Device 5 according to Fig. 5.5c. The first quantum well also yields the stronger radiative recombination rate in spite of the low electron concentration [see Fig. 5.5b] for Device 5, which is because of the suppressed polarization induced electric field intensity as demonstrated in Fig. 5.6. The suppressed polarization induced electric

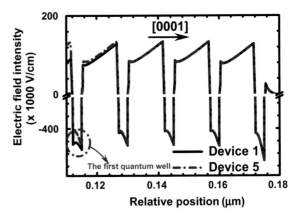

Fig. 5.6 Electric field profiles in the MQWs for selectively chosen Devices 1 and 5, respectively at the current density of 160 A/cm^2

field in the first quantum well arises from the smaller polarization mismatch between the n-Al$_{0.52}$Ga$_{0.38}$N layer and the Al$_{0.55}$Ga$_{0.45}$N quantum barrier. The polarization induced electric field in the other quantum well seems not to be affected by the n-Al$_{0.52}$Ga$_{0.38}$N underneath layer. Hence the promoted radiative recombination in Fig. 5.5c is uniquely ascribed to the increased electron concentration level.

To summarize, we have proven that the electron injection can be enhanced by "cooling down" the free electrons, i.e., the electron energy decreases. We propose to reduce the electron drift velocity and the electron energy by using stair-cased AlN composition for the n-AlGaN electron injection layer and/or modulating the Si doping concentration in the n-AlGaN layer (i.e., EFR). By using the stair-cased AlN composition for the n-AlGaN layer, we are able to obtain the polarization induced electric field at each n-Al$_x$Ga$_{1-x}$N/n-Al$_y$Ga$_{1-y}$N interface. The polarization induced electric field is opposite to the external-bias-generated electric field, hence "slowing down" the electrons. The reduced electron energy by using the modulated Si doping concentration in the n-AlGaN layer utilizes the build-in electric field at the n-Al$_x$Ga$_{1-x}$N/n-Al$_y$Ga$_{1-y}$N interface, for which the Si doping concentration for the n-Al$_y$Ga$_{1-y}$N layer is lower than that for the n-Al$_x$Ga$_{1-x}$N layer. The build-in electric field also opposes to the external-bias-generated electric field, which therefore also reduces the electron drift velocity. Our study shows that the electron concentration and the EQE for the proposed structures are improved. Nevertheless, if the Si doping concentration is not properly set, the proposed structure may increase the forward voltage, which correspondingly sacrifices the wall-plug-efficiency. Therefore, the thickness for the n-Al$_y$Ga$_{1-y}$N layer with the modulated Si doping concentration shall be fully optimized. Our report on the EFR provides the alternative design freedom for improving the electron injection efficiency of AlGaN based DUV LEDs. In addition, our report is the make-up work for the existing device physics for AlGaN based DUV LEDs.

References

1. Zhang Z-H, Zhang Y, Bi W, Demir HV, Sun XW (2016) On the internal quantum efficiency for InGaN/GaN light-emitting diodes grown on insulating substrates. Phys Status Solidi (a) 213(12):3078–3102. https://doi.org/10.1002/pssa.201600281
2. Cho J, Schubert EF, Kim JK (2013) Efficiency droop in light-emitting diodes: challenges and countermeasures. Laser Photonics Rev 7(3):408–421. https://doi.org/10.1002/lpor.201200025
3. Sun W, Shatalov M, Deng J, Hu X, Yang J, Lunev A, Bilenko Y, Shur M, Gaska R (2010) Efficiency droop in 245–247 nm AlGaN light-emitting diodes with continuous wave 2 mW output power. Appl Phys Lett 96(6):061102. https://doi.org/10.1063/1.3302466
4. Zhang Z-H, Chen S-WH, Chu C, Tian K, Fang M, Zhang Y, Bi W, Kuo H-C (2018) Nearly efficiency-droop-free AlGaN-based ultraviolet light-emitting diodes with a specifically designed superlattice p-type electron blocking layer for high Mg doping efficiency. Nanoscale Res Lett 13:122. https://doi.org/10.1186/s11671-018-2539-9
5. Huang J, Guo ZY, Guo M, Liu Y, Yao SY, Sun J, Sun HQ (2017) Study of deep ultraviolet light-emitting diodes with a p-AlInN/AlGaN superlattice electron-blocking layer. J Electron Mater 46(7):4527–4531. https://doi.org/10.1007/s11664-017-5413-0
6. Kuo Y-K, Chen F-M, Lin B-C, Chang J-Y, Shih Y-H, Kuo H-C (2016) Simulation and experimental study on barrier thickness of superlattice electron blocking nayer in near-ultraviolet light-emitting diodes. IEEE J Quantum Electron 52(8):1–6. https://doi.org/10.1109/JQE.2016.2587100
7. Sun P, Bao XL, Liu SQ, Ye CY, Yuan ZR, Wu YK, Li SP, Kang JY (2015) Advantages of AlGaN-based deep ultraviolet light-emitting diodes with a superlattice electron blocking layer. Superlattices Microstruct 85:59–66. https://doi.org/10.1016/j.spmi.2015.05.010
8. Zhang Z-H, Chen S-WH, Zhang Y, Li L, Wang S-W, Tian K, Chu C, Fang M, Kuo H-C, Bi W (2017) Hole transport manipulation to improve the hole injection for deep ultraviolet light-emitting diodes. Acs Photonics 4(7):1846–1850. https://doi.org/10.1021/acsphotonics.7b00443
9. Chu CS, Tian KK, Fang MQ, Zhang YH, Li LP, Bi WG, Zhang ZH (2018) On the $Al_xGa_{1-x}N/Al_yGa_{1-y}N/Al_xGa_{1-x}N$ (x > y) p-electron blocking layer to improve the hole injection for AlGaN based deep ultraviolet light-emitting diodes. Superlattices Microstruct 113:472–477. https://doi.org/10.1016/j.spmi.2017.11.029
10. Chu C, Tian K, Fang M, Zhang Y, Zhao S, Bi W, Zhang Z-H (2018) Structural design and optimization of deep-ultraviolet light-emitting diodes with $Al_xGa_{1-x}N/Al_yGa_{1-y}N/AlxGa_{1-x}N$ (x > y) p-electron blocking layer. J Nanophotonics 12(4):043503, May 2018, https://doi.org/10.1117/1.jnp.12.043503
11. So B, Kim J, Shin E, Kwak T, Kim T, Nam O (2018) Efficiency improvement of deep-ultraviolet light emitting diodes with gradient electron blocking layers. Physica Status Solidi a-Appl Mater Sci 215(10):1700677. https://doi.org/10.1002/pssa.201700677
12. Kwon MR, Park TH, Lee TH, Lee BR, Kim TG (2018) Improving the performance of AlGaN-based deep-ultraviolet light-emitting diodes using electron blocking layer with a heart-shaped graded Al composition. Superlattices Microstruct 116:215–220. https://doi.org/10.1016/j.spmi.2018.02.033
13. Chen Q, Zhang J, Geo Y, Chen JW, Long HL, Dai JN, Zhang ZH, Chen CQ (2018) Improved the AlGaN-based ultraviolet LEDs prformance with super-lattice structure last barrier. IEEE Photonics J 10(4):1–7. https://doi.org/10.1109/JPHOT.2018.2852660
14. Tian K, Chen Q, Chu C, Fang M, Li L, Zhang Y, Bi W, Chen C, Zhang Z-H, Dai J (2018) Investigations on AlGaN-based deep-ultraviolet light-emitting diodes with Si-doped quantum barriers of different doping concentrations. Physica Status Solidi-Rapid Res Lett 12(1):1700346. https://doi.org/10.1002/pssr.201700346
15. Chang JY, Chang HT, Shih YH, Chen FM, Huang MF, Kuo YK (2017) Efficient carrier confinement in deep-ultraviolet light-emitting diodes with composition-graded configuration. IEEE Trans Electron Devices 64(12):4980–4984. https://doi.org/10.1109/TED.2017.2761404

16. Zhang Z-H, Chu C, Chiu CH, Lu TC, Li L, Zhang Y, Tian K, Fang M, Sun Q, Kuo H-C, Bi W (2017) UVA light-emitting diode grown on Si substrate with enhanced electron and hole injections. Opt Lett 42(21):4533–4536. https://doi.org/10.1364/OL.42.004533

17. Zhang Z-H, Tian K, Chu C, Fang M, Zhang Y, Bi W, Kuo H-C (2018) Establishment of the relationship between the electron energy and the electron injection for AlGaN based ultraviolet light-emitting diodes. Opt Express 26(14):17977–17987. https://doi.org/10.1364/oe.26.017977

18. Fang M, Tian K, Chu C, Zhang Y, Zhang Z-H, Bi W (2018) Manipulation of Si doping concentration for modification of the electric field and carrier injection for AlGaN-based deep-ultraviolet light-emitting diodes. Crystals 8(6):258. https://doi.org/10.3390/cryst8060258

19. Li L, Shi Q, Tian K, Chu C, Fang M, Meng R, Zhang Y, Zhang Z-H, Bi W (2017) A dielectric-constant-controlled tunnel junction for III-nitride light-emitting diodes. Physica Status Solidi a-Appl Mater Sci 214(6):1600937. https://doi.org/10.1002/pssa.201600937

20. Li L, Zhang Y, Tian K, Chu C, Fang M, Meng R, Shi Q, Zhang Z-H, Bi W (2017) Numerical investigations on the n^+-GaN/AlGaN/p^+-GaN tunnel junction for III-nitride UV light-emitting diodes. Physica Status Solidi a-Appl Mater Sci 214(12):1700624. https://doi.org/10.1002/pssa.201700624

Chapter 6
Screen the Polarization Induced Electric Field Within the MQWs for DUV LEDs

Abstract This chapter discusses and presents different designs to screen the polarization level in the quantum wells for [0001]-oriented DUV LEDs. By doing so, the quantum confined Stark effect (QCSE) can be decreased. We suggest a simple way to reduce the QCSE by adopting Si-doped quantum barriers. Meanwhile, we also find that DUV LEDs are very sensitive to the polarization polarity, such that if nonpolar, semipolar and nitrogen-polar DUV LED structures are grown, we shall avoid using the p-AlGaN/p-GaN hole injection layer. The p-AlGaN/p-GaN hole injection layer can have remarkably hole depletion effect at the interface for those growth orientations except the [0001] orientation.

After the carriers are injected into the MQW region, the radiative recombination between electrons and holes produces DUV photons. At the current stage, DUV LEDs grown along [0001] orientation still take the dominant place both in research and industry. [0001] oriented AlGaN/AlGaN MQWs possess very strong polarization induced electric field, which spatially separates electron and hole wave functions, known as quantum confined Stark effect (QCSE) [1]. A most convenient method to suppress the polarization induced electric field is doping the quantum barriers by Si dopants [2]. In Fig. 6.1, we calculate and then show the optical power and the wave function overlap in terms of the Si doping concentration in the quantum barriers for the DUV LEDs. We can get that the overlap for the wave functions of electrons and holes increases as more Si dopants are contained in the quantum barriers. The increased overlap for the carrier wave functions originates from the free electrons that are released by the Si dopants. The free carriers can compensate the polarization induced charges at the interface for the quantum barrier/quantum well pairs. Nevertheless, the optical power does not monotonically increase with the increasing overlap for carriers according to Fig. 6.1, which is believed to be caused by the hole blocking effect when the Si doping concentration for the quantum barriers becomes high. Detailed discussions will be conducted subsequently. The inset for Fig. 6.1 shows the numerically calculated and the experimentally measured forward voltage in terms of the Si doping concentration in the quantum barriers. The forward voltage is probed when the current density is 30 A/cm^2. We can get that the increase of the Si doping concentration in the quantum barriers reduces the forward voltage both

Z.-H. Zhang et al., *Deep Ultraviolet LEDs*, Nanoscience and Nanotechnology,
https://doi.org/10.1007/978-981-13-6179-1_6

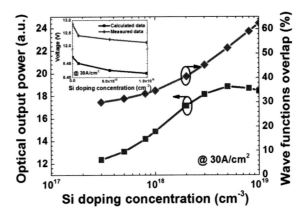

Fig. 6.1 Numerically calculated optical power and wave functions overlap for the investigated DUV LEDs in terms of the Si doping concentration in the quantum barriers. Inset show the numerically calculated and the experimentally measured forward voltage as a function of the Si doping concentration in the quantum barriers. The data are calculated and measured at the current density of 30 A/cm^2. Reproduced from Ref. [2], with the permission of Wiley Publishing

experimentally and numerically. Although the similar trending for the relationship between the forward voltage and the Si doping concentration in the quantum barriers is obtained, the experimentally measured forward voltage is larger than the numerical ones, and this is likely to be caused by the unoptimized process for ohmic contacts. However, the observations here do not affect our conclusions for the comparative study.

To further study the impact of the Si doped quantum barriers on the polarization effect and the hole injection, we selectively grow four DUV LEDs by using metal organic chemical vapor deposition (MOCVD) technique, which are Devices R, I, II and III. Device R has undoped quantum barriers, while the Si doping concentrations are ~8 × 10^{17}, ~5 × 10^{18} and ~1 × 10^{19} cm^{-3} for Devices I, II and III respectively. We next present the electric field profiles, hole concentration profiles and electron concentration profiles in Fig. 6.2a, b and c, respectively. Figure 6.2a shows that the electric field intensity in the quantum wells is reduced when the quantum barriers are doped with Si. Meanwhile, with the increasing Si doping level, the electric field intensity further decreases, meaning that the polarization induced electric field in the quantum wells can be screened by doping the quantum barrier with Si dopants for AlGaN based DUV LEDs. We then investigate the influence of different Si doping concentrations in the quantum barriers on the hole injection efficiency across the MQW region [see Fig. 6.2b]. From Fig. 6.2b we can get that the hole concentration in the quantum wells apart from the p-EBL is the highest for Device R among the investigated devices. The holes are more accumulated in the quantum wells close to the p-EBL when the quantum barriers are doped with Si dopants. We can also see that the strongest hole accumulation occurs at the last quantum well for Devices II and III. Here, the last quantum well is defined as the quantum well closest to the p-EBL. The hole blocking effect by the Si-doped quantum barriers will be explained subsequently by using Fig. 6.3. Figure 6.2c shows the electron concentration levels for Devices

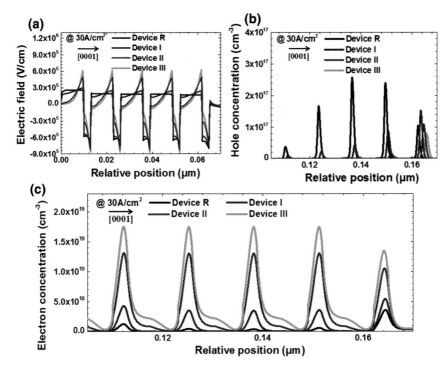

Fig. 6.2 **a** Electric field profiles, **b** hole concentration profiles and **c** electron concentration profiles across the MQWs for Devices R, I, II and III, respectively. Figures are calculated at the current density of 30 A/cm². Reproduced from Ref. [2], with the permission of Wiley Publishing

R, I, II and III, respectively. We can conclude that, by doping the quantum barriers with Si dopants, the electron concentration in the quantum wells can be significantly enhanced. The Si dopants in the quantum barriers replenish the electrons that escape from the MQW region.

As has been mentioned previously [see Fig. 6.2b], the hole injection is remarkably hindered when the quantum barriers are doped with Si dopant. Then, we show the energy band diagrams for Devices R, I, II and III in Fig. 6.3a, b, c and d, respectively. If we compare the valence band of the quantum barriers between Devices R and III, we can see that the valence band of the quantum barriers for Device III is bent in the way of being concave. The observed "curvature" in the valence band for Device III is attributed to the ionized Si dopants, which gives rise to the "depletion effect" and hence increasing the valence band barrier height. Such "depletion effect" will become more obvious when the Si doping level increases. Table 6.1 summarizes the effective valence band barrier height ($\Delta\Phi$) of the quantum barriers for Devices R, I, II and III. We can clearly see that the values for the valence band barrier height increases with the increasing Si doping concentration level in the quantum barriers. Figure 6.3a–d and Table 6.1 well interprets the results in Figs. 6.1 and 6.2b.

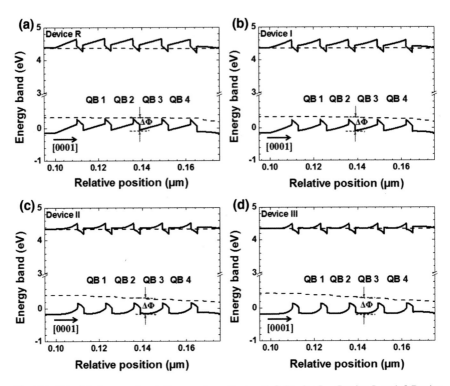

Fig. 6.3 Calculated energy band digrams for **a** Device R, **b** Device I, **c** Device I, and **d** Device III. QB1, QB2, QB3 and QB3 represent quantum barrier 1, quantum barrier 2, quantum barrier 3 and quantum barrier 4, respectively. $\Delta\Phi$ means the effective valence band barrier height for holes. Figures are calculated at the current density of 30 A/cm². Reproduced from Ref. [2], with the permission of Wiley Publishing

Table 6.1 Values of $\Delta\Phi$ in the quantum barriers for Devices R, I, II and III. Values are calculated at the current density of 30 A/cm²

	Device R ($\Delta\Phi$) (meV)	Device I ($\Delta\Phi$)	Device II ($\Delta\Phi$)	Device III ($\Delta\Phi$)
QB 1	385	437	537	583
QB 2	376	424	495	528
QB 3	376	409	452	472
QB 4	378	394	408	417

Reproduced from Ref. [2], with the permission of Wiley Publishing

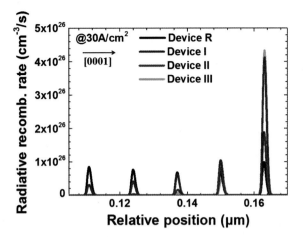

Fig. 6.4 Numerically calculated radiative recombination rate in the MQWs for Devices R, I, II and III. Data are calculated at the current density of 30 A/cm². Reproduced from Ref. [2], with the permission of Wiley Publishing

The Si doping concentration affects the polarization induced electric field, the electron concentration and the hole concentration in the MQW region. As the interplay for the polarization effect, the electron concentration and the hole concentration, it is worth studying the radiative recombination rate for the four devices [see Fig. 6.4]. The radiative recombination rate profiles are consistent with the hole concentration distributions in Fig. 6.2b. The homogeneous hole distribution in the MQW region for Device R takes account for the uniform radiative recombination rate. The radiative recombination rate is strongly accumulated in the last quantum well for Devices II and III.

The overall radiative recombination rate is reflected by the EL intensity and the optical power. Figure 6.5a presents the experimentally measured EL spectra for Devices R, I, II and III when the current density is 30 A/cm². We can see that the EL intensity increases if we look into Devices R, I and II. However, the EL intensity decreases for Device III which has the Si doping concentration of 1×10^{19} cm^{-3} in the quantum barriers. The evolutional details for the measured EL intensity is consistent with our calculated results in Fig. 6.1 such that the EL intensity does not monotonically increases with the Si doping concentration in the quantum barriers. Further analysis into Fig. 6.5a shows the blue shift for the peak emission wavelength for Devices I, II and III when compared to Device R. The blue shift for the peak emission wavelength is well attributed to the polarization screening effect by the Si dopant. However, further blue shift of the wavelength with the increasing Si doping concentration in the quantum barriers is not obviously observed for Devices I, II, and III, and this might be caused by the slightly different AlN compositions in the Al-rich AlGaN quantum wells from run to run during the epitaxial growth process. Figure 6.5b shows the numerically calculated EL spectra for Devices R, I, II and III when the current density is 30 A/cm². The relationship between calculated EL intensity and the Si doping concentration in the quantum barriers agrees very well with the one in Fig. 6.5a. Meanwhile, we also see the blue shift for the peak emission wavelength when the quantum barriers are doped with Si dopants. The

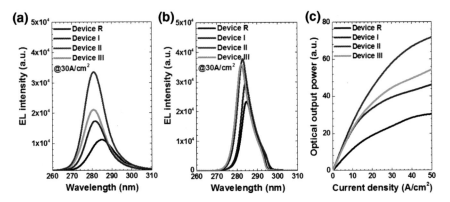

Fig. 6.5 **a** Experimentally measured EL spectra, **b** numerically calculated EL spectra, **c** experimentally measured optical power in terms of the injection current density level for Devices R, I, II and III, respectively. Fig. **a** and **b** are measured and calculated at the current density of 30 A/cm^2, respectively. Reproduced from Ref. [2], with the permission of Wiley Publishing

excellent agreement between Fig. 6.5a and Fig 6.5b provides the solid evidence that the physical models and parameters are properly set in our calculation. After measuring the EL spectra for the tested DUV LEDs, we are able to get the optical output power as a function of the injection current density for Devices R, I, II and III in Fig. 6.5c. The results shown in Fig. 6.5c are also consistent with our calculations in Fig. 6.1.

In a summary, our work regarding the Si-doped quantum barriers shows that, by using the Si-doped quantum barriers, the polarization induced electric field in the MQWs for the [0001] oriented DUV LEDs can be screened. Moreover, the screening capability can be even more enhanced if more Si dopants can be incorporated into the quantum barriers and are ionized to provide free electrons. The Si-doped quantum barriers also improve the electron concentration in the active region and reduce the forward voltage. Nevertheless, we also find that we shall be very careful with the hole injection if the Si-doped quantum barriers are adopted to DUV LEDs. If the Si dosage level is too high, the hole injection can be significantly blocked, which leads to a degraded EQE. It means that the EQE and the optical power are not always monotonically increasing with the Si doping concentration in the quantum barriers. Besides optimizing the Si doping level in the quantum barriers, we also suggest optimizing the number for the quantum barriers with Si dopants incorporated, such that two sets of quantum well/quantum barrier architectures can be designed in the active region. One set of the quantum well/quantum barrier structure is to promote hole injection while the other set of quantum well/quantum barrier structures is to screen the polarization effect for high electron-hole radiative recombination rate.

Besides the common methods are proposed to screen the polarization induced electric field in the MQW region, other alternatives are also demonstrated. Li et al. demonstrate that the polarization induced electric field in the quantum well becomes weak when n-AlGaN underlying layer is grown before the MQWs stack [3]. They also suggest that the AlN composition for the n-AlGaN underlying layer is lower than that for the n-AlGaN template. Most recently, semipolar AlGaN based DUV LEDs have been reported, and the epitaxial growth is initiated on the r-plane sapphire substrate [4]. Although the results for Ref. [4] do not mention whether the QCSE has been decreased, the stable wavelength and narrow emission line width with the increasing injection current level indicate the suppressed QCSE. Therefore, the semipolar or the nonpolar AlGaN/AlGaN quantum well structures provide another design freedom to high-efficiency DUV LEDs. Methods for suppressing the QCSE have been reported for InGaN/GaN visible LEDs [5, 6], and we believe most of the approaches are also doable for DUV LEDs. However, when designing nonpolar and semipolar AlGaN based active region, one shall avoid using the p-AlGaN/p-GaN hole injection layer. Note, according to the findings by our group, we require the p-AlGaN/p-GaN hole injection layer to be along the [0001] orientation with very strong polarization effect to suppress the hole depletion at the p-AlGaN/p-GaN junction [7]. Otherwise, the hole depletion effect at the p-AlGaN/p-GaN interface can reduce the hole injection. Therefore, we suggest not using the p-AlGaN/p-GaN structure for nonpolar and semipolar DUV LEDs. Detailed discussions will be conducted subsequently.

In our work, the semipolar, metal-polar, nitrogen-polar and nonpolar features are represented by setting the polarization level, which reflects the crystalline relaxation degree and the polarization charge density in this work, e.g., if the polarization level of 0.4 is assumed, then 0.6 of the theoretical polarization charges are released by generating dislocations. Note, the polarization level of 1 indicates the fully strained epitaxial layer without relaxation. Here, the polarization charges are induced by the coupled effect of the spontaneous and the piezoelectric polarizations. The polarization level of zero means nonpolar devices. Note, some semipolar oriented III-nitride films are also free from the polarization effect, for which case the polarization level is also set to zero. The positive polarization level (e.g., 0.4) and the negative polarization level (e.g., -0.4) represents the metal-polar/semipolar and the nitrogen-polar/semipolar devices, respectively.

The investigated DUV LEDs possess the 4 μm thick Si doped $Al_{0.60}Ga_{0.40}N$ layer in which the electron concentration is 5×10^{18} cm^{-3}. Then, the architectural stack comprising five pairs of $Al_{0.45}Ga_{0.55}N$ (3 nm)/$Al_{0.57}Ga_{0.43}N$ (10 nm) MQWs serves as the active region, which enables the peak emission wavelength of ~280 nm. The following layer is the 10 nm thick Mg doped $Al_{0.60}Ga_{0.40}N$ p-EBL. Next, the p-type hole injection layer is a heterojunction consisting of 50 nm thick $Al_{0.40}Ga_{0.60}N$ layer and 50 nm thick GaN layer. The hole concentration for the Mg doped layers is set to 3×10^{17} cm^{-3}. The mesa size is set to 350×350 μm. To better investigate the impact of polarization effect of the p-EBL/p-$Al_{0.40}Ga_{0.60}N$/p-GaN structure on the performance for DUV LEDs, we also selectively choose the devices with various polarization levels at different layers for detailed discussions on the hole concentration, the energy band alignment and the electric field profiles [see Table 6.2].

Figure 6.6a shows that the light output power (LOP) is associated with the polarization level. With the polarization level varying from -1 to 1, the LOP increases. Interestingly, the LOP for [000-1] oriented DUV LEDs is lower than that for [0001] oriented DUV LEDs. Moreover, for [0001] oriented DUV LEDs, the LOP is enhanced

Table 6.2 Structure information for the studied devices

Devices	Layer position with varied polarization	Polarization level				
		$i = 1$	$i = 2$	$i = 3$	$i = 4$	$i = 5$
Ai	Whole structure	−0.8	−0.4	0	0.4	0.8
Bi	p-Al$_{0.40}$Ga$_{0.60}$N/p-GaN interface	−0.2	0	0.2	0.6	×
Ci	p-EBL/p-Al$_{0.40}$Ga$_{0.60}$N interface	−0.2	0	0.2	0.6	×
Di	p-EBL/p-Al$_{0.40}$Ga$_{0.60}$N and p-Al$_{0.40}$Ga$_{0.60}$N/p-GaN interfaces	−0.2	0	0.2	0.6	×

'i' represents the device number. Data are calculated at the current of 35 mA. Reproduced from Ref. [7], with the permission of Springer Publishing

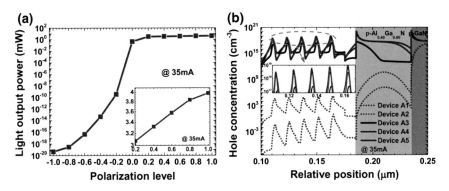

Fig. 6.6 **a** Light output power in terms of the polarization level varying from −1 to 1, **b** hole concentration profiles for Devices A1, A2, A3, A4 and A5 with different polarization levels. Data are calculated at the current of 35 mA. Reproduced from Ref. [7], with the permission of Springer Publishing

with the increasing polarization level. The conclusions drawn here are quite contrary to the common belief such that the stronger polarization effect reduces the electron-hole radiative recombination rate within the MQWs. However, it is also known that the electron-hole radiative recombination rate is also co-affected by the hole injection. Figure 6.6b illustrates the hole concentration profiles for Devices A1, A2, A3, A4 and A5 with different polarization levels of −0.8, −0.4, 0, 0.4 and 0.8 at the current of 35 mA, respectively. As the polarization level increases from −0.8 to 0.8, the hole concentration in the MQWs increases. The very low hole concentration in the [000-1] oriented MQWs well interprets the poor LOP in Fig. 6.6a. Figure 6.6b also presents the hole concentration profiles in the p-Al$_{0.40}$Ga$_{0.60}$N/p-GaN region, and we can see that the hole concentration in the p-Al$_{0.40}$Ga$_{0.60}$N layer becomes lower once the DUV LED is [000-1] oriented. The hole concentration in the p-Al$_{0.40}$Ga$_{0.60}$N layer can be enhanced when the DUV LED is grown along the [0001] orientation. Hence, we believe the polarization level of the p-EBL/p-Al$_{0.40}$Ga$_{0.60}$N and p-Al$_{0.40}$Ga$_{0.60}$N/p-GaN interfaces remarkably influence the hole injection, which is to be conducted subsequently.

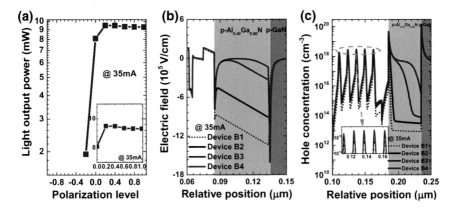

Fig. 6.7 **a** Light output power in terms of the polarization level varying from −0.2 to 1 for p-Al$_{0.40}$Ga$_{0.60}$N/p-GaN interface, **b** electric field profiles, and **c** hole concentration profiles for Devices B1, B2, B3 and B4, respectively. Data are calculated at the current of 35 mA. Reproduced from Ref. [7], with the permission of Springer Publishing

Firstly, we vary the polarization level of the p-Al$_{0.40}$Ga$_{0.60}$N/p-GaN interface and keep the polarization of 0.4 for other layers, since by doing so, the variation of the LOP can be uniquely attributed to the different hole injections. Figure 6.7a shows the LOP variation with different polarization levels of the p-Al$_{0.40}$Ga$_{0.60}$N/p-GaN interface at the current of 35 mA, which shows that the LOP is dramatically improved once the polarity for the p-Al$_{0.40}$Ga$_{0.60}$N/p-GaN interface turns to the [0001] polarity. Furthermore, the LOP is not strongly affected if the polarization level of the p-Al$_{0.40}$Ga$_{0.60}$N/p-GaN interface further increases from 0 to 1. To better illustrate the origin for the observations in Fig. 6.7a, we selectively choose Devices B1, B2, B3, and B4 with the polarization levels for the p-Al$_{0.40}$Ga$_{0.60}$N/p-GaN interface of −0.2, 0, 0.2 and 0.6 for further analysis, respectively. Figure 6.7b and c show the electric field and hole concentration profiles for the four studied devices at the current of 35 mA, respectively. Figure 6.7b depicts that the electric field for Devices B1 and B2 is very strong throughout the entire p-Al$_{0.40}$Ga$_{0.60}$N layer and Device B1 has strongest electric field intensity. As the polarization level increases to 0.2 and 0.6, respectively, the electric field intensity for Devices B3 and B4 decreases. According to the report by us, the electric field can facilitate the hole injection by adjusting the drift velocity and the kinetic energy for holes [8]. Thus, we further calculate the values for the work done to holes by the electric field within p-Al$_{0.40}$Ga$_{0.60}$N/GaN structure for Devices B1, B2, B3 and B4, which are summarized in Table 6.3. As is illustrated, holes in Device B1 can obtain the most energy among the studied devices, and holes can obtain the least energy from Device B4. Figure 6.7c shows that, with the polarization level varying from −0.2 to 0.6, the hole concentration in the p-Al$_{0.40}$Ga$_{0.60}$N layer increases, e.g., the holes in the p-Al$_{0.40}$Ga$_{0.60}$N layer for Device B1 is the lowest, which translates the smallest hole concentration in the MQW region though the holes receive the largest energy from the p-Al$_{0.40}$Ga$_{0.60}$N/p-GaN region [see Table 6.3].

Table 6.3 Values of the work done to holes by the electric field within the p-Al$_{0.40}$Ga$_{0.60}$N/p-GaN structure for Devices B1, B2, B3 and B4, '−' represents that the holes obtain energy from the electric field

Devices	Device B1	Device B2	Device B3	Device B4
Work (meV)	−5605	−3556	−965	−614

Data are calculated at the current of 35 mA. Reproduced from Ref. [7], with the permission of Springer Publishing

The significant hole depletion in the p-Al$_{0.40}$Ga$_{0.60}$N layer for Device B2 also hinders the hole injection into the MQW in spite of the very large energy that the holes obtain. Device B3 and B4 have the higher hole concentration level in the MQWs according to Fig. 6.7c thanks to the less hole depletion in the p-Al$_{0.40}$Ga$_{0.60}$N layer. Note, the hole concentrations for Device B3 and B4 within the MQWs are quite similar because of the compromise between the hole concentration in the p-Al$_{0.40}$Ga$_{0.60}$N layer and the energy that the holes obtain from the p-Al$_{0.40}$Ga$_{0.60}$N layer.

Figure 6.8 demonstrates the energy band diagrams for Devices B1, B2, B3 and B4 at the current of 35 mA. It is known that the potential barrier height (φ) at p-Al$_{0.40}$Ga$_{0.60}$N/GaN interface can hinder the holes injection into p-Al$_{0.40}$Ga$_{0.60}$N layer leading to hole depletion in the p-Al$_{0.40}$Ga$_{0.60}$N layer. Clearly, we can see that the values of φ for Devices B1, B2, B3 and B4 are 399, 366, 342 and 248 meV, respectively. Hence, the hole concentration in the p-Al$_{0.40}$Ga$_{0.60}$N layer for Device B1 is significantly depleted. Although Device B1 provides holes with the largest energy, the very low hole concentration in the p-Al$_{0.40}$Ga$_{0.60}$N layer accounts for the lowest hole concentration in the MQWs and the poorest LOP for Device B1 [see Fig. 6.7a and c]. The same physical principle also applies to Device B2. Device B3 shows the slightly better performance than Device B4, since it is a compromise of both the enhanced energy for holes and the decent hole concentration in the p-Al$_{0.40}$Ga$_{0.60}$N layer [see Table 6.3 and Fig. 6.7c]. Thus, it is summarized that the improved performance for DUV LEDs can be realized as long as the polarization level for the p-Al$_{0.40}$Ga$_{0.60}$N/GaN interface is moderately adjusted. Note, when the polarization level is lower than −0.2 in Fig. 6.7a, the DUV LEDs have more severer hole depletion in p-Al$_{0.40}$Ga$_{0.60}$N layer, and thus result in the insufficient hole injection that causes the numerical nonconvergence when conducting calculations.

Next, we discuss the impact of the polarization level variation at the p-EBL/p-Al$_{0.40}$Ga$_{0.60}$N interface on the performance for DUV LEDs. The polarization level of 0.4 is set for the other layers. Figure 6.9a presents the LOP in terms of different polarization levels at the current of 35 mA. Obviously, the LOP tendency is similar to the obtained one in Fig. 6.6a. For insightful discussions, we choose Devices C1, C2, C3 and C4, for which the polarization levels of the p-EBL/p-Al$_{0.40}$Ga$_{0.60}$N interface are −0.2, 0, 0.2 and 0.6 respectively. The electric field and hole concentration profiles for the four devices are demonstrated in Fig. 6.9b and c. According to Fig. 6.9b, the electric field for Device C1 is significantly different from others due to the positive polarization charges at the p-EBL/p-Al$_{0.40}$Ga$_{0.60}$N interface. Figure 6.9b also reveals

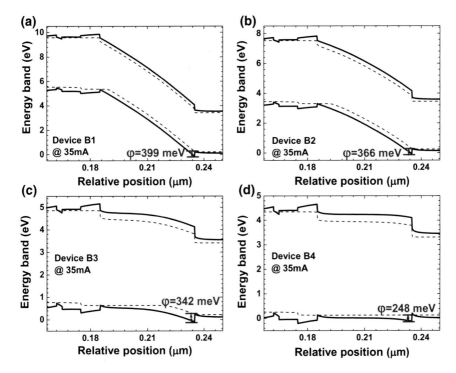

Fig. 6.8 Energy band diagrams for **a** Device B1, **b** Device B2, **c** Device B3, **d** Device B4. The polarization level for the p-Al$_{0.40}$Ga$_{0.60}$N/p-GaN interface is set to -0.2, 0, 0.2 and 0.6, respectively at the current of 35 mA. φ represents the valence band barrier height for holes at p-Al$_{0.40}$Ga$_{0.60}$N/p-GaN interface. Data are calculated at the current of 35 mA. Reproduced from Ref. [7], with the permission of Springer Publishing

the electric field intensity for other devices increases with the increasing polarization level, i.e. Device C2 < Device C3 < Device C4. The calculated values of the work done to holes by the electric field within p-Al$_{0.40}$Ga$_{0.60}$N/GaN structures for Devices C1, C2, C3 and C4 are summarized in Table 6.4. Holes can obtain less energy from Device C1 once the p-EBL/p-Al$_{0.40}$Ga$_{0.60}$N interface is of the [000-1] polarity, and the energy of holes for Devices C2, C3 and C4 increases as the polarization level becomes large. Figure 6.9c demonstrates the hole concentration profiles for Devices C1, C2, C3 and C4. The hole concentration in the MQWs for Devices C3 and C4 is higher than that for Device C2 with the hole concentration being the highest for Device C4, and the hole concentration in the MQWs for Device C1 is the lowest.

Figure 6.10 presents the energy band diagrams for Devices C1, C2, C3 and C4 at the current of 35 mA. The hole concentration in the MQWs is both affected by the hole concentration in the p-Al$_{0.40}$Ga$_{0.60}$N layer and the valence band barrier height for the p-EBL (Φ). As is illustrated, the hole concentration for Device C1 is severely depleted in p-Al$_{0.40}$Ga$_{0.60}$N layer closing to p-EBL and the Φ reaches 792 meV due to the positive polarization charges at p-EBL/p-Al$_{0.40}$Ga$_{0.60}$N interface. It accounts

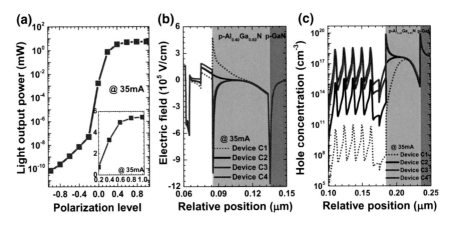

Fig. 6.9 **a** Light output power in terms of the polarization level varying from −1 to 1 for p-EBL/p-$Al_{0.40}Ga_{0.60}N$ interface, **b** electric field profiles, and **c** hole concentration profiles for Devices C1, C2, C3 and C4, respectively. Data are calculated at the current of 35 mA. Data are calculated at the current of 35 mA. Reproduced from Ref. [7], with the permission of Springer Publishing

Table 6.4 Values of the work done to holes by the electric field within the p-$Al_{0.40}Ga_{0.60}N$/p-GaN structure for Devices C1, C2, C3 and C4, '−' represents that the holes obtain energy from the electric field

Devices	Device C1	Device C2	Device C3	Device C4
Work (meV)	−67	−382	−532	−636

Data are calculated at the current of 35 mA. Data are calculated at the current of 35 mA. Reproduced from Ref. [7], with the permission of Springer Publishing

for the lowest hole concentration in the MQWs and the poorest LOP for Device C1 [see Fig. 6.9a and c]. Once the p-EBL/p-$Al_{0.40}Ga_{0.60}N$ interface turns to the [0001] polarity, the hole accumulation at the interface of p-EBL/p-$Al_{0.40}Ga_{0.60}N$ causes a reduced value of Φ [9], i.e., the values of Φ are 529, 390 and 351 meV for Devices C2, C3 and C4, respectively. As a result, holes can efficiently transport and inject into the active region especially Device C3 and C4, which interprets the results shown in Fig. 6.9a and c. Hence, it is concluded that the improved performance for DUV LEDs can be obtained by making the p-EBL/p-$Al_{0.40}Ga_{0.60}N$ interface [0001] oriented and keeping the increased polarization level.

Finally, we simultaneously vary the polarization level of p-EBL/p-$Al_{0.40}Ga_{0.60}N$ and p-$Al_{0.40}Ga_{0.60}N$/GaN interfaces while setting the polarization level of 0.4 for the other layers. Figure 6.11a tells that the LOP increases as the polarization level increases from −1 to 1, which is consistent with the results observed in Fig. 6.6a. Next, we show the electric filed and hole concentration profiles in Fig. 6.6b and c for Devices D1, D2, D3 and D4 for which the polarization levels of the p-EBL/p-$Al_{0.40}Ga_{0.60}N$ and p-$Al_{0.40}Ga_{0.60}N$/GaN interfaces are −0.2, 0, 0.2 and 0.6, respectively. Moreover, the values for the work done to holes by the electric field within the

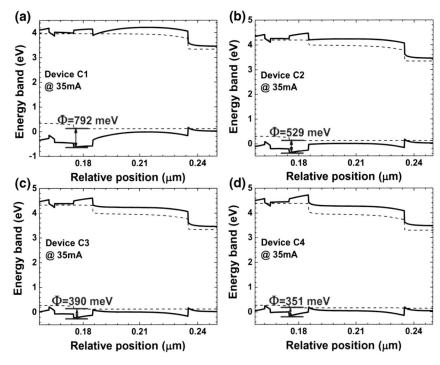

Fig. 6.10 Energy band diagrams for **a** Device C1, **b** Device C2, **c** Device C3, **d** Device C4. The polarization level for the p-EBL/p-Al$_{0.40}$Ga$_{0.60}$N interface is set to −0.2, 0, 0.2 and 0.6, respectively at the current of 35 mA. Φ represents the valence band barrier height for p-EBL. Data are calculated at the current of 35 mA. Reproduced from Ref. [7], with the permission of Springer Publishing

p-Al$_{0.40}$Ga$_{0.60}$N/p-GaN structure are also calculated and summarized in Table 6.5. Figure 6.6b demonstrates that partial of electric field in the p-Al$_{0.40}$Ga$_{0.60}$N/p-GaN structure for Device D1 is opposite to the hole injection path. As a result, the holes may be slowed down and retarded by the p-Al$_{0.40}$Ga$_{0.60}$N layer and p-EBL. The electric field for Devices D2, D3 and D4 is along the hole injection path and can facilitate the hole injection into the MQWs by enhancing energy for holes, which is well confirmed by Table 6.5. Meanwhile, the previous analysis also reveals that the valence band barrier height for the p-EBL can be significantly reduced once more holes are accumulated at the p-EBL/p-Al$_{0.40}$Ga$_{0.60}$N interface [see Fig. 6.10]. Hence, the hole concentration in the MQWs for Devices D2, D3, and D4 is far higher than that for Device D1, which is shown in Fig. 6.11c. Figure 6.11c also illustrates the highest hole concentration is given by Device D4, which possesses the largest kinetic energy for holes and smallest valence band barrier height of the p-EBL. Hence, we verify that the polarization level of the p-EBL/p-Al$_{0.40}$Ga$_{0.60}$N and p-Al$_{0.40}$Ga$_{0.60}$N/p-GaN interfaces remarkably influence the performance for DUV LEDs. We can intentionally increase the polarization level of p-AlGaN layer to improve the hole injection efficiency for DUV LEDs. Meanwhile, we notice the LOP for Device D4 is ~5 mW

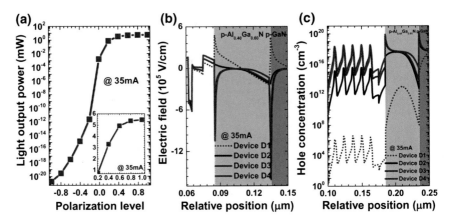

Fig. 6.11 **a** Light output power in terms of the polarization level varying from −1 to 1 for p-EBL/p-Al$_{0.40}$Ga$_{0.60}$N and p-Al$_{0.40}$Ga$_{0.60}$N/p-GaN interfaces, **b** electric field profiles, and **c** hole concentration profiles for Devices D1, D2, D3 and D4, respectively. Data are calculated at the current of 35 mA. Reproduced from Ref. [7], with the permission of Springer Publishing

Table 6.5 Values of the work done to holes by the electric field within the p-Al$_{0.40}$Ga$_{0.60}$N/p-GaN structure for Devices D1, D2, D3 and D4, '−' represents the holes obtain energy from the electric field

Devices	Device 1	Device 2	Device 3	Device 4
Work (meV)	+236	−323	−526	−652

Data are calculated at the current of 35 mA. Reproduced from Ref. [7], with the permission of Springer Publishing

[see Fig. 6.11a], which number is higher than that of ~3.6 mW in Fig. 6.6a when the polarization degree is 0.6 for the whole DUV LED architecture. The difference of the LOP arises from the different polarization levels in the active region. Hence, we strongly believe the performance can be further boosted by screening the polarization field in the MQWs.

In summary, the report in Ref. [7] reveals that the polarity for the p-EBL/p-AlGaN/p-GaN hole injection layer is essentially important for DUV LEDs. We suggest not using the conventional p-EBL/p-AlGaN/p-GaN hole injection layer when one tries to grow semipolar, nonpolar and nitrogen-polar DUV LEDs. The hole injection can be facilitated only when the p-EBL/p-AlGaN/p-GaN hole injection layer is of the metal polarity. Meanwhile, the strong polarization level for the [0001] oriented p-EBL/p-AlGaN/p-GaN hole injection layer is also required, which enables the suppression of the hole depletion effect in the p-AlGaN layer. The results here are also consistent with our findings in Ref. [10]. Therefore, we strongly believe that more efforts shall be made for the in-depth investigations into the p-EBL/p-AlGaN/p-GaN hole injection layer for AlGaN based DUV LEDs. We also suggest the alternative hole injection layer when growing semipolar, nonpolar and nitrogen-polar DUV LEDs, e.g., using the p-AlGaN based hole injection layer with the AlN composition graded.

References

1. Li X, Sundaram S, Disseix P, Le Gac G, Bouchoule S, Patriarche G, Reveret F, Leymarie J, El Gmili Y, Moudakir T, Genty F, Salvestrini JP, Dupuis RD, Voss PL, Ougazzaden A (2015) AlGaN-based MQWs grown on a thick relaxed AlGaN buffer on AlN templates emitting at 285 nm. Opt Mater Express 5(2):380–392. https://doi.org/10.1364/ome.5.000380

2. Tian K, Chen Q, Chu C, Fang M, Li L, Zhang Y, Bi W, Chen C, Zhang Z-H, Dai J (2018) Investigations on AlGaN-based deep-ultraviolet light-emitting diodes with Si-doped quantum barriers of different doping concentrations. Physica Status Solidi-Rapid Res Lett 12(1):1700346. https://doi.org/10.1002/pssr.201700346

3. Li L, Miyachi Y, Miyoshi M, Egawa T (2016) Enhanced emission efficiency of deep ultraviolet light-emitting AlGaN multiple quantum wells grown on an n-AlGaN underlying layer. IEEE Photonics J 8(5):1–10. https://doi.org/10.1109/JPHOT.2016.2601439

4. Akaike R, Ichikawa S, Funato M, Kawakami Y (2018) $Al_xGa_{1-x}N$-based semipolar deep ultraviolet light-emitting diodes. Appl Phys Express 11, Jun 2018

5. Zhang Z-H, Zhang Y, Bi W, Demir HV, Sun XW (2016) On the internal quantum efficiency for InGaN/GaN light-emitting diodes grown on insulating substrates. Phys Status Solidi (a) 213(12):3078–3102. https://doi.org/10.1002/pssa.201600281

6. Wang L, Jin J, Mi C, Hao Z, Luo Y, Sun C, Han Y, Xiong B, Wang J, Li H (2017) A review on experimental measurements for understanding efficiency droop in InGaN-based light-emitting diodes. Materials 10(11):1233. https://doi.org/10.3390/ma10111233

7. Tian K, Chu C, Shao H, Che J, Kou J, Fang M, Zhang Y, Bi W, Zhang Z-H (2018) On the polarization effect of the p-EBL/p-AlGaN/p-GaN structure for AlGaN-based deep-ultraviolet light-emitting diodes. Superlattices Microstruct 122:280–285. https://doi.org/10.1016/j.spmi.2018.07.037

8. Zhang Z-H, Li L, Zhang Y, Xu F, Shi Q, Shen B, Bi W (2017) On the electric-field reservoir for III-nitride based deep ultraviolet light-emitting diodes. Opt Express 25(14):16550–16559. https://doi.org/10.1364/OE.25.016550

9. Zhang Z-H, Chen S-WH, Zhang Y, Li L, Wang S-W, Tian K, Chu C, Fang M, Kuo H-C, Bi W (2017) Hole transport manipulation to improve the hole injection for deep ultraviolet light-emitting diodes. ACS Photonics 4(7):1846–1850. https://doi.org/10.1021/acsphotonics.7b00443

10. Zhang Z-H, Zhang Y, Bi W, Geng C, Xu S, Demir HV, Sun XW (2016) On the hole accelerator for III-nitride light-emitting diodes. Appl Phys Lett 108(15):071101. https://doi.org/10.1063/1.4947025

Chapter 7
Thermal Management for DUV LEDs

Abstract DUV LEDs possess very huge heating issue. On one hand, the sapphire substrate has a poor thermal conductivity, and on the other hand, DUV photons are easily absorbed by the absorptive p-GaN layer and the metal contact in the way of free carrier absorption, which further increases the self-heating effect for DUV LEDs. This chapter briefs readers on the currently adopted technologies for better thermal management.

DUV LEDs emit photons with very high energy which can be easily absorbed by those non-active layers. Meanwhile, the low LEE causes very strong free carrier absorption. Therefore, those DUV photons that fail to escape from the chips are converted into heat. On the other hand, sapphire has very low thermal conductivity $(0.35 \text{ W cm}^{-1} \text{ K}^{-1})$, which further makes the thermal management important for DUV LEDs. As has been mentioned previously, the self-heating can be reduced by improving the current spreading effect [1]. Guo et al. suggest driving the DUV LEDs by using the pulse mode with the duty cycle smaller than 0.2% [2]. However, most of the solid-state light sources are driving by CW power source. Thus, DUV LEDs are featured with the flip-chip structure and the devices are mounted on the AlN insulation carrier [3]. Note, the AlN insulation carrier can well dissipate the heat because of the excellent thermal conductivity of $175 \text{ W cm}^{-1} \text{ K}^{-1}$.

© The Author(s), under exclusive license to Springer Nature Singapore Pte Ltd. 2019   59
Z.-H. Zhang et al., *Deep Ultraviolet LEDs*, Nanoscience and Nanotechnology,
https://doi.org/10.1007/978-981-13-6179-1_7

References

1. Hao GD, Taniguchi M, Tamari N, Inoue S (2018) Current crowding and self-heating effects in AlGaN-based flip-chip deep-ultraviolet light-emitting diodes. J Phys D-Applied Phys 51(3):035103. https://doi.org/10.1088/1361-6463/aa9e0e
2. Guo H, Yang Y, Cao XA (2008) Thermal and nonthermal factors affecting the quantum efficiency of deep-ultraviolet light-emitting diodes. Phys Status Solidi a-Appl Mater Sci 205(12):2953–2957. https://doi.org/10.1002/pssa.200824046
3. Shatalov M, Chitnis A, Yadav P, Hasan MF, Khan J, Adivarahan V, Maruska HP, Sun WH, Khan MA (2005) Thermal analysis of flip-chip packaged 280 nm nitride-based deep ultraviolet light-emitting diodes. Appl Phys Lett 86(20):201109. https://doi.org/10.1063/1.1927695

Chapter 8
The Light Extraction Efficiency for DUV LEDs

Abstract DUV LEDs have very low light extraction efficiency (LEE), which is caused by the unique optical polarization and the optically absorptive semiconductor and metal layers. This chapter reviews and analyzes the approaches that have ever been used to improve the LEE. This chapter also points out that, the removal of the p-GaN layer can yield a high LEE without guaranteeing the enhanced wall plug efficiency in the same time. Thus, even more effort shall be made to achieve excellent ohmic contact for DUV LEDs.

DUV LEDs are unique because of the co-existence of the TE-polarized and TM-polarized photons in the MQW region. Therefore, the LEE for both the TE-polarized and TM-polarized photons has to be enhanced.

First of all, the LEE is subject to the optical absorption by the p-GaN layer that has the energy bandgap lower than the AlGaN based MQWs. According to the report by Shatalov et al. [1], the LEE can be improved by over 50% if the p-GaN layer can be replaced by the UV transparent Al-rich p-AlGaN layer. However, the difficulty in obtaining high hole concentration for Al-rich p-AlGaN layer makes the perfect ohmic Al reflector less doable. Fortunately, Pd has the work function of 5.12 eV that is larger than Al of 3.28 eV. Moreover, the electric resistivity for Pd is smaller than 10^{-3} Ω cm^{-2}, which makes Pd as the excellent candidate for p-type ohmic current spreading layer. Nevertheless, the optical reflectivity for Pd is as low as 40% in the UV spectral range. Therefore, Lobo eta al. propose to utilize nanopixel contacts on the p-AlGaN layer [2], such that the nanopixel Pd contacts are patterned and fabricated on the p-AlGaN layer for DUV LEDs, and then the Al reflector is deposited on both the Pd nanopixel contacts and the p-AlGaN layer [see Fig. 8.1a]. By doing so, the LEE for DUV LEDs can be increased without significantly sacrificing the forward voltage as along as the size for the nanopixels are fully optimized [see Fig. 8.1b]. Another approach to relieve the optical absorption is to form Ni/p-GaN meshes by conducting selectively etching [3], and then Al reflector is deposited on the Ni/p-GaN and the exposed p-AlGaN layer [see Fig. 8.1c]. By doing so, the LEE can be enhanced by 1.55 while the forward voltage is negligibly increased as shown in Fig. 8.1d.

Another effective approach for increasing the LEE without increasing the forward voltage is sample patterning. The DUV LEDs can be grown on the patterned

Z.-H. Zhang et al., *Deep Ultraviolet LEDs*, Nanoscience and Nanotechnology,
https://doi.org/10.1007/978-981-13-6179-1_8

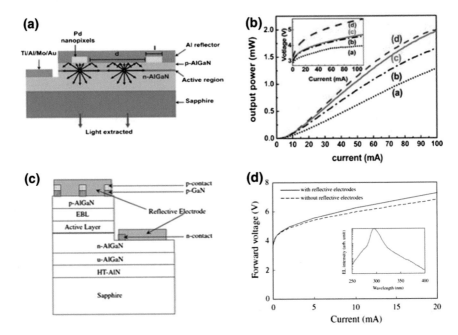

Fig. 8.1 a Schematic cross-sectional view for the DUV LED with Pd contacts and Al reflector layer, **b** optical output power in terms of the injection current for nanopixel DUV LEDs with different nanopixel sizes; inset shows the relationship between the current and the voltage for the structures, **c** schematic cross-sectional view for the DUV LED with the proposed p-GaN/Ni/Au ohmic contacts and Al reflector layer, **d** current-voltage characteristics for DUV LEDs with and without reflective electrodes; inset shows the emission spectrum at 20 mA DC operation Fig. **a** and **b** reproduced from Ref. [2], with the permission of AIP Publishing; Fig. **c** and **d** reproduced from Ref. [3], with permission of IOP Publishing

substrates, and the light escape surface can be patterned, e.g., the backside for the sapphire substrate. It has been proven that growing InGaN/GaN LEDs on patterned sapphire substrate can at least double the LEE [4]. The patterns on the substrate are required to be in nanoscale which is because of the low mobility for Al adatoms and the long coalescence time when growing AlN buffer layer [5–8]. Besides engineering the substrate, patterning the backside for the device is an effective alternative approach to increase the LEE for flip-chip DUV LEDs, e.g., hybridized structures comprising AlN nanocone photonics crystals and the subwavelength nanostructures on the AlN substrate [9], nanolens arrays [10], microlens [11], textured the N-face AlN buffer layer that is wet-etched in the KOH solution [12].

As has been mentioned previously, the optical emission for DUV LEDs is dominated by the TM-polarized photons, which indicates that the majority of the light will propagate to the sidewall of the mesa. Therefore, it is essentially important to increase the LEE on the device sidewalls. For that purpose, roughened sidewalls for the sapphire are fabricated by using picosecond laser dicing [13], which are shown

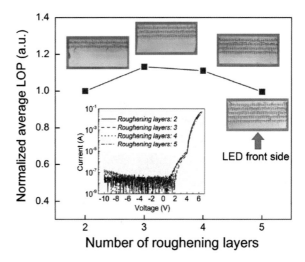

Fig. 8.2 Simulated LEE enhancement factor as a function of the roughening layer number. Reproduced from Ref. [13], with the permission of AIP Publishing

in Fig. 8.2. We can also see that the LEE reaches the maximum when the roughened layer is 3. The inset for Fig. 8.2 also indicates that the roughen sidewalls for the sapphire substrate do not affect the current-voltage characteristics, since the sapphire serves as the insulator without providing current paths.

Another design to enhance the LEE for the TM-polarized light is to modulate the propagation direction of the beams, which can be achieved by adopting the inclined side wall reflector [14–17]. The beam propagation can be tuned and the LEE for DUV LEDs can also be enhanced by using the inclined mesa [18, 19]. The optimized inclined angle for the mesa is found to be 37.83° according to the report by Chen et al. [19]. Most recently, our group has further modified the inclined sidewall structure, such that we place the bottom metal mirror with air cavities in the adjacent mesas so that the TM-polarized DUV photons can be less absorbed by the sidewall metal mirrors [20, 21].

Since most of the TM-polarized light tends to escape from the sidewall, then it will be realistic to increase the surface-to-volume ratio by fabricating nanorod LEDs. The other advantage for nanorod UV LEDs is the partial release of the strain [22]. According to numerical calculations, the LED for both TE- and TM-polarized light can be remarkably increased as presented in Fig. 8.3a and b [23]. The LEE for nanorod DUV LEDs can be further promoted if the proper passive layer is deposited on the nanorods, and the numerical calculations indicate that, among the candidates of SiO_2, AlN and SiN_x, the small refractive index of SiO_2 can best extract the photons from the nanorod cores [24]. In spite of the promising aspect for nanorod DUV LEDs, the fabrication difficulty is a big obstacle, and the surface damages that are caused during the dry etching process for the top-to-down nanorod structures make the EQE even lower than 0.01% at the current stage [25].

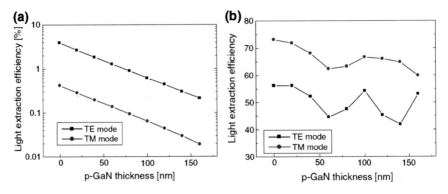

Fig. 8.3 Numerically calculated LEE of TE- and TM-polarized emission in terms of the p-GaN layer thickness for **a** planar DUV LEDs, **b** nanorod DUV LEDs. Reproduced from Ref. [23], with the permission of Springer

References

1. Shatalov M, Sun W, Jain R, Lunev A, Hu X, Dobrinsky A, Bilenko Y, Yang J, Garrett GA, Rodak LE, Wraback M, Shur M, Gaska R (2014) High power AlGaN ultraviolet light emitters. Semicond Sci Technol 29(8):084007. https://doi.org/10.1088/0268-1242/29/8/084007
2. Lobo N, Rodriguez H, Knauer A, Hoppe M, Einfeldt S, Vogt P, Weyers M, Kneissl M (2010) Enhancement of light extraction in ultraviolet light-emitting diodes using nanopixel contact design with Al reflector. Appl Phys Lett 96(8):081109. https://doi.org/10.1063/1.3334721
3. Inazu T, Fukahori S, Pernot C, Kim MH, Fujita T, Nagasawa Y, Hirano A, Ippommatsu M, Iwaya M, Takeuchi T, Kamiyama S, Yamaguchi M, Honda Y, Amano H, Akasaki I (2011) Improvement of light extraction efficiency for AlGaN-based deep ultraviolet light-emitting diodes. Jpn J Appl Phys 50:122101. https://doi.org/10.1143/jjap.50.122101
4. Zhmakin AI (2011) Enhancement of light extraction from light emitting diodes. Phys Rep-Rev Sect Phys Lett 498(4–5):189–241. https://doi.org/10.1016/j.physrep.2010.11.001
5. Kim M, Fujita T, Fukahori S, Inazu T, Pernot C, Nagasawa Y, Hirano A, Ippommatsu M, Iwaya M, Takeuchi T, Kamiyama S, Yamaguchi M, Honda Y, Amano H, Akasaki I (2011) AlGaN-based deep ultraviolet light-emitting diodes fabricated on patterned sapphire substrates. Appl Phys Express 4(9):092102. https://doi.org/10.1143/APEX.4.092102
6. Dong P, Yan J, Wang J, Zhang Y, Geng C, Wei T, Cong P, Zhang Y, Zeng J, Tian Y, Sun L, Yan Q, Li J, Fan S, Qin Z (2013) 282-nm AlGaN-based deep ultraviolet light-emitting diodes with improved performance on nano-patterned sapphire substrates. Appl Phys Lett 102(24):241113. https://doi.org/10.1063/1.4812237
7. Adivarahan V, Fareed Q, Islam M, Katona T, Krishnan B, Khan A (2007) Robust 290 nm emission light emitting diodes over pulsed laterally overgrown AlN. Jpn J Appl Phys Part 2-Lett & Express Lett 46(36):L877–L879, Oct 2007. https://doi.org/10.1143/jjap.46.l877
8. Lee D, Lee JW, Jang J, Shin IS, Jin L, Park JH, Kim J, Lee J, Noh HS, Kim YI, Park Y, Lee GD, Kim JK, Yoon E (2017) Improved performance of AlGaN-based deep ultraviolet light-emitting diodes with nano-patterned AlN/sapphire substrates. Appl Phys Lett 110(19):191103. https://doi.org/10.1063/1.4983283
9. Inoue S-I, Naoki T, Kinoshita T, Obata T, Yanagi H (2015) Light extraction enhancement of 265 nm deep-ultraviolet light-emitting diodes with over 90 mW output power via an AlN hybrid nanostructure. Appl Phys Lett 106(13):131104. https://doi.org/10.1063/1.4915255

10. Liang R, Dai J, Xu L, He J, Wang S, Peng Y, Wang H, Ye L, Chen C (2018) High light extraction efficiency of deep ultraviolet LEDs enhanced using nanolens arrays. IEEE Trans Electron Devices 65(6):2498–2503. https://doi.org/10.1109/ted.2018.2823742
11. Khizar M, Fan ZY, Kim KH, Lin JY, Jiang HX (2005) Nitride deep-ultraviolet light-emitting diodes with microlens array. Appl Phys Lett 86(17):173504. https://doi.org/10.1063/1.1914960
12. Lachab M, Asif F, Zhang B, Ahmad I, Heidari A, Fareed Q, Adivarahan V, Khan A (2013) Enhancement of light extraction efficiency in sub-300 nm nitride thin-film flip-chip light-emitting diodes. Solid-State Electron 89:156–160. https://doi.org/10.1016/j.sse.2013.07.010
13. Guo YA, Zhang Y, Yan JC, Xie HZ, Liu L, Chen X, Hou MJ, Qin ZX, Wang JX, Li JM (2017) Light extraction enhancement of AlGaN-based ultraviolet light-emitting diodes by substrate sidewall roughening. Appl Phys Lett 111(1):011102. https://doi.org/10.1063/1.4991664
14. Wierer JJ Jr, Allerman AA, Montano I, Moseley MW (2014) Influence of optical polarization on the improvement of light extraction efficiency from reflective scattering structures in AlGaN ultraviolet light-emitting diodes. Appl Phys Lett 105(6):061106. https://doi.org/10.1063/1.4892974
15. Kim DY, Park JH, Lee JW, Hwang S, Oh SJ, Kim J, Sone C, Schubert EF, Kim JK (2015) Overcoming the fundamental light-extraction efficiency limitations of deep ultraviolet light-emitting diodes by utilizing transverse-magnetic-dominant emission. Light-Sci Appl 4(4):e263–e263. https://doi.org/10.1038/lsa.2015.36
16. Lee JW, Park JH, Kim DY, Schubert EF, Kim J, Lee J, Kim YI, Park Y, Kim JK (2016) Arrays of truncated cone AlGaN deep-ultraviolet light-emitting diodes facilitating efficient outcoupling of in-plane emission. Acs Photonics 3(11):2030–2034. https://doi.org/10.1021/acsphotonics.6b00572
17. Lee JW, Kim DY, Park JH, Schubert EF, Kim J, Lee J, Kim YI, Park Y, Kim JK (2016) An elegant route to overcome fundamentally-limited light extraction in AlGaN deep-ultraviolet light-emitting diodes: Preferential outcoupling of strong in-plane emission. Sci Rep 6(1):22537. https://doi.org/10.1038/srep22537
18. Guo YN, Zhang Y, Yan JC, Chen X, Zhang S, Xie HZ, Liu P, Zhu HF, Wang JX, Li JM (2017) Sapphire substrate sidewall shaping of deep ultraviolet light-emitting diodes by picosecond laser multiple scribing. Appl Phys Express 10(6):062101. https://doi.org/10.7567/APEX.10.062101
19. Chen Q, Zhang H, Dai J, Zhang S, Wang S, He J, Liang R, Zhang Z-H, Chen C (2018) Enhanced the optical power of AlGaN-based deep ultraviolet light-emitting diode by optimizing mesa sidewall angle. IEEE Photonics J 10(4):6100807. https://doi.org/10.1109/JPHOT.2018.2850038
20. Zhang Y, Meng R, Zhang Z-H, Shi Q, Li L, Liu G, Bi W (2017) Effects of inclined sidewall structure with bottom metal air cavity on the light extraction efficiency for AlGaN-based deep ultraviolet light-emitting diodes. IEEE Photonics J 9(5):1600709. https://doi.org/10.1109/JPHOT.2017.2736642
21. Zhang Y, Zheng Y, Meng R, Sun C, Tian K, Geng C, Zhang Z-H, Liu G, Bi W (2018) Enhancing both TM- and TE-polarized light extraction efficiency of AlGaN-based deep ultraviolet light-emitting diode via air cavity extractor with vertical sidewall. IEEE Photonics J 10(4):8200809. https://doi.org/10.1109/JPHOT.2018.2849747
22. Dong P, Yan JC, Zhang Y, Wang JX, Geng C, Zheng HY, Wei XC, Yan QF, Li JM (2014) Optical properties of nanopillar AlGaN/GaN MQWs for ultraviolet light-emitting diodes. Opt Express 22(5):A320–A327. https://doi.org/10.1364/OE.22.00A320
23. Ryu HY (2014) Large enhancement of light extraction efficiency in AlGaN-based nanorod ultraviolet light-emitting diode structures. Nanoscale Res Lett 9(1):58. https://doi.org/10.1186/1556-276X-9-58
24. Ooi YK, Liu C, Zhang J (2017) Analysis of polarization-dependent light extraction and effect of passivation layer for 230-nm AlGaN nanowire light-emitting diodes. IEEE Photonics J 9(4):4501712. https://doi.org/10.1109/JPHOT.2017.2710325
25. Zhao S, Sadaf SM, Vanka S, Wang Y, Rashid R, Mi Z (2016) Sub-milliwatt AlGaN nanowire tunnel junction deep ultraviolet light emitting diodes on silicon operating at 242 nm. Appl Phys Lett 109(20):325. https://doi.org/10.1063/1.4967837

Chapter 9
Conclusions and Outlook

Abstract This chapter summarizes the content for this book and suggests the future research outlooks for DUV LEDs.

To summarize, this work has reviewed the issues that AlGaN-based DUV LEDs are encountering now. Countermeasures that are taken by the community are reported. The absence of the localized states makes the IQE for DUV LEDs sensitive to the TDs, and the TDD can be reduced by growing DUV LEDs on nanopatterned substrates. Because of the poor Mg doping efficiency for the p-AlGaN layer, the low electrical conductivity easily causes the current crowding effect for the flip-chip DUV LEDs. Nevertheless, the current spreading for DUV LEDs are less studied at the current stage, and this can be suggested as a research spot. To increase the EQE for DUV LEDs, one has to also increase the hole injection, which can be realized by improving the p-type doping efficiency for the p-type nitride layer, increasing the hole drift velocity, reducing the hole blocking effect by the p-EBL and increasing the hole concentration within the MQWs. It is interesting that the hole distribution across the MQWs for DUV LEDs shows various profiles, and this is caused by the different energy band offset ratios assumed during the calculations. Moreover, the $Al_xGa_{1-x}N/Al_yGa_{1-y}N$ quantum well/quantum barrier structure provides more freedom in setting the AlN compositions (i.e., x and y), which also affect the hole distribution profiles across the MQW region. We also summarize the designs which are effective in reducing the electron leakage. Specifically, we present to reduce the electron energy by modulating the Si doping concentration or/and the AlN composition in the n-AlGaN layer. The quantum wells can better capture the electrons once the electron energy decreases. This work discusses the importance of screening the polarization effect within the MQWs and the excellent thermal management for DUV LEDs. Lastly, it is pointed out that the LEE for DUV LEDs can be improved by e.g., using nanopatterns on the backside for the sapphire, depositing reflective mirrors on the sidewalls for the mesa, adopting the inclined mesa sidewall. Thanks to the

Z.-H. Zhang et al., *Deep Ultraviolet LEDs*, Nanoscience and Nanotechnology,
https://doi.org/10.1007/978-981-13-6179-1_9

larger surface-to-volume ratio, nanorod DUV LEDs are also promising for increase the LEE numerically, though the EQE for nanorod DUV LEDs is low experimentally. The authors in this work also suggest investigating the in-depth device physics for DUV LEDs by using advanced simulation program. The understanding for the device physics is important for developing high-efficiency DUV LEDs at a low cost.

Appendix

In all our calculations, Investigations on the DUV LEDs are made by using Crosslight APSYS. We set the Shockley-Read-Hall (SRH) lifetime and Auger recombination coefficient to 10 ns and 1×10^{-30} cm^6 s^{-1} in our model, respectively [1,2]. The energy band offset ratio between the conduction band offset and the valence band offset for AlGaN/AlGaN heterojunctions is 50:50 [3]. The polarization charges can be calculated by referring to the method developed by Fiorentini et al. [4]. The light extraction efficiency for the studied DUV LEDs is $\sim 9\%$ which is within the reasonable range for the ~ 280 nm DUV photons [5].

References

1. Zhang Z-H, Li L, Zhang Y, Xu F, Shi Q, Shen B, Bi W (2017) On the electric-field reservoir for III-nitride based deep ultraviolet light-emitting diodes. Opt Express 25(14):16550–16559. https://doi.org/10.1364/OE.25.016550
2. Tian K, Chen Q, Chu C, Fang M, Li L, Zhang Y, Bi W, Chen C, Zhang Z-H, Dai J (2018) Investigations on AlGaN-based deep-ultraviolet light-emitting diodeswith Si-doped quantum-barriers of different doping concentrations. Physica Status Solidi-Rapid Res Lett 12(1):1700346. https://doi.org/10.1002/pssr.201700346
3. Piprek J (2010) Phys Status Solidi A 207 (2010), 2217–2225
4. Fiorentini V et al (2002) Appl Phys Lett 80:1024
5. Ryu H-Y, Choi I-G, Choi H-S, Shim J-I (2013) Investigation of light extraction efficiency in AlGaN deep-ultraviolet light-emitting diodes. Appl Phys Express 6(6):062101. https://doi.org/10.7567/APEX.6.062101

Printed in the United States
By Bookmasters